A Resilience Approach to Acceleration of Sustainable Development Goals

Mika Shimizu

Editor

A Resilience Approach to Acceleration of Sustainable Development Goals

 Springer

Editor
Mika Shimizu
Graduate School of Advanced Integrated
Studies in Human Survivability
Kyoto University
Kyoto, Japan

ISBN 978-981-19-4347-8   ISBN 978-981-19-4345-4 (eBook)
https://doi.org/10.1007/978-981-19-4345-4

This Springer imprint is published by the registered company Springer Nature Singapore Pte Ltd.
The registered company address is: 152 Beach Road, #21-01/04 Gateway East, Singapore 189721,
Singapore

*Dedicated to Dr. Allen L. Clark (1938–2021)
Thank you for guiding me through my
research life to get to this point. You have sent
me a great gift. I will pass it on to the next
generation.*

Mika Shimizu

May 2022

Preface

From a broad perspective, the resilience approach to sustainable development goals is to promote transdisciplinary research given the approach faces complex and systemic problems through diverse perspectives across time and scales by linking different types of knowledge through discourse toward addressing a common good.

Through the resilience approach which focuses on linkages and boundaries in nature-human-society systems or living systems, it attempts to bridge the global to local scales, sustainability to disaster management, and policy-science-communities, where many missing links exist, often preventing the implementation of SDGs.

Identifying missing links cannot be done automatically, and it is difficult to identify them in a fixed group or committee of members, or through a fixed process and scale. The resilience approach demonstrates it is imperative for us to review our society through different axis to gain new insights and synergies to seek viable solutions. In other words, humans need to make the most of our creativity and imagination which cannot be replaced by the machine.

Kyoto, Japan   Mika Shimizu

Acknowledgements

The lead author and editor of the book would like to express special thank you to Dr. Norio Okada, Emeritus Professor, Kyoto University, Dr. Ilan Chabay, Head of Strategic Science Initiatives and Senior Investigator in the Real Deal European Commission project at the Institute for Advanced Sustainability Studies (IASS), Dr. Kaoru Takara, Emeritus Professor, Kyoto University, and Dr. Toshiyuki Yasui, Dean of the Department of Social System Design of Eikei University of Hiroshima, who have provided valuable comments on the topics in this book or provided great supports for the relevant research activities.

This work was partly supported by JST SICORP Grant Number JPMJSC2117, Japan.

Contents

Part IV   Educational Perspectives

Part V   Conclusion

Editor and Contributors

About the Editor

Mika Shimizu is an Associate Professor in Graduate School of Advanced Integrated Studies in Human Survivability, Kyoto University. Her long years' experiences as a policy researcher in East-West Center in Washington DC and Honolulu, Hawaii in the United States greatly contributed to publishing this book. She holds an M.A. from American University and a Ph.D. in International Public Policy from Osaka University (2006). She has been extensively involved in interdisciplinary and trans-disciplinary research projects related to disasters/infectious diseases, sustainability, and climate change issues with the focus on resilience. Her major publications include *Nexus of Resilience and Public Policy in a Modern Risk Society* (Co-Author: Allen Clark, Springer, 2019).

Contributors

Chabay Ilan Institute for Advanced Sustainability Studies (IASS), Potsdam, Germany

Nakamura Hidenori Department of Environmental and Civil Engineering, Toyama Prefectural University, Imizu, Japan

Noguchi Fumiko United Nations University Institute for the Advanced Study of Sustainability (UNU-IAS), Tokyo, Japan

Okada Norio Kwansei Gakuin University, Nishinomiya, Japan

Shimizu Mika Graduate School of Advanced Integrated Studies in Human Surviv-ability, Kyoto University, Kyoto, Japan;
Kyoto University, Kyoto, Japan

Tong Vincent C. H. Department of Geography and Environmental Sciences, Northumbria University, Newcastle upon Tyne, UK

Part I
A Resilience Approach for Sustainable Development Goals (SDGs)

Chapter 1
Introduction: Transformation for Systemic Challenges

Mika Shimizu

Abstract This chapter, as an introductory chapter of this book, provides the book scope and backgrounds including why this book focuses on a resilience approach for Sustainable Development Goals (SDGs), and how the approach is interrelated with systemic challenges in and around SDGs across natural-human-social systems (living systems). Especially, this chapter highlights the underlying theme of this book, i.e., the transformation for systemic challenges, which is the major concerns of SDGs at local through global scales. Thus, the chapter articulates the structure of systemic challenges, such as complex cause-effect structures from macro to microlevels, multiple interacting components, multidimensional and cascading impacts and "deep uncertainty," and discusses how the natures of systemic challenges are relevant to a resilience approach. With this backbone, this chapter introduces different chapters in this book.

1.1 About the Book

1.1.1 Introduction

A resilience approach to acceleration of sustainable development goals (SDGs) provides pathways for implementing SDGs through a resilience approach. The resilience approach directs a society's challenges toward problem-solving-oriented actions in operationalizing resilience, such as nurturing, strengthening, or creating resilience, in different layers of human societies in harmony with nature against adversaries. A resilience approach to the novel coronavirus (COVID-19) does not mean returning to the 2019 status quo; however, building a capacity of learning and designing the adaptive capacity of systems to deal with future shocks and stress (Trump et al., 2021).

M. Shimizu (✉)
Graduate School of Advanced Integrated Studies in Human Survivability, Kyoto University, Kyoto, Japan

Why does this book focus on a resilience approach? The reason is systemic challenges driven by natural, human, and social risks at the local through global levels, many of which are integrated into SDGs by the United Nations with the target at 2015–2030. The characteristics of systemic challenges include (1) complex cause-effect structures from macro to microlevels, (2) multiple interacting components, and (3) multidimensional and cascading impacts and deep uncertainties (see the details in Sect. 1.2.1). The predecessor of SDGs, Millennium Development Goals (MDGs) (2000–2015), target developing countries, while challenges in and around SDGs require collective actions with developing and developed countries and all kind of stakeholders, including businesses given systemic challenges at different scales and dimensions.

Another reason, which is related to the first one, is that humans, ecology, and communities/societies are supposed to inherently embrace capacities to change and recover, i.e., resilience; however, those capacities can be protected, nurtured, and strengthened only if they maintain interdependent "relationships" of nature-human–social systems and keep options open and emphasize heterogeneity (Holling, 1973). Thus, the focus on the common trait of resilience and their essential common nature of "relationships" be a key in overcoming the systemic challenges.

Given the systemic challenges our society faces, not **fixing "a dot" but "connecting dots to form lines" approach is essential in addressing the challenges. This is an aspect of a resilience approach**. If we miss connecting dots, **the missing link may result in "tradeoffs" between SDGs, i.e., an action that looks appropriate in promoting one of SDGs may impact negatively on another aspect of SDGs**. For example, suppose you construct solar panels in forests by cutting trees for energy efficiency or climate change without looking at surroundings around the construction site, such as human or wildlife living areas adjacent to the construction sites. In that case, the construction may affect the living conditions of the humans and wildlife, leading to disasters in human living areas when severe typhoons or floods occur, or it may affect biodiversity because of natural environment destruction by construction. That is, tradeoffs between SDG #7 (clean energy) or #13 (climate change) and #11 (sustainable cities) and # 15 (life on land) may occur (see more discussions in Chapter 9).

A special attention area highlighted through a resilience approach is *linkages* and *boundaries. The linkages and boundaries* in diverse contexts and scales will hold a key in *enabling resilience* and problem-solving-oriented actions (Shimizu & Clark, 2019). One of the major issues the resilience approach addresses is "silo" problems between/among disciplines, systems/subsystems, sectors, organizations, and issues, which often become barriers against enabling resilience, by looking at the pieces and whole in a continuum as well, especially with the attention to linkages and boundaries. This book attempts to step forward toward better approaches and actions for SDGs by providing a relevant conceptual framework and providing case studies or practices that are linked, mislinked, or dis-linked with the resilience approach.

1.1.2 Objective and Background

The objective of this book is to (1) identify operational gaps and missing links in implementing SDGs, (2) demonstrate how approaches to systemic challenges in and around SDGs can impact the trajectory of actions, processes in actions, and in turn, outcomes from actions, and (3) provide pathways for addressing systemic challenges in and around SDGs or natural-human–social systems for transformation toward acceleration for SDGs. While SDGs are well known at different levels from local to global spheres, major gaps are seen between goals and approaches: Though the United Nations General Assembly in 2015 acknowledged interlinkages as being of crucial importance in ensuring the purpose of the goals, the current approaches tend to focus on separate goals in SDGs or separate systems in natural spheres such as sea, forest, and land, social dimensions and individual needs, and rights except for a few emerging efforts (see Chapter 2). However, systemic challenges dynamically affected our society (see Sect. 1.2) which requires systemic and synergetic actions to create viable solutions by stakeholders with the focus on natural-human-social system linkages (see Chapter 2).

As a current state of implementing SDGs, few specific approaches addressed the interlinkages or interconnections through policy, organizational, community, and practical levels (see Chapter 2). Grainger-Brown and Malekpour (2019) found that while many tools and frameworks for implementing SDGs have been developed to support organizations in engaging with SDGs, most of the tools are only applicable to mapping. and reporting activities. Few tools exist which engaged with actual strategy development for transformation (see the detail assessment in Chapter 2).

Approaches matter because it provides a framework for addressing challenges, which affects social orientation and outcomes from collective actions. SDGs articulate-multifaceted problems reflecting our complex and uncertain age, where natural, social, and human challenges originate from human activities and are interlinked at diverse scales and dimensions. How to approach the challenges influences all types of processes, including learning, capacity-building, and management/governance at community, project, program, organizational, or social levels, which will affect outcomes and consequences by human actions. Approaches affect how to put resources and capacities on a trajectory toward problem-solving-oriented actions.

A resilience approach represents not a single uniform to fix problems, but a synthesis of resilience thinking, resilience-driven actionable knowledge, and practices, including systems approach (see Chapter 2), to navigate the systemic challenges or dynamics of natural-human–social systems. As such, a resilience approach provides navigation for problem-solving-oriented actions with alternatives or options by looking at both the contexts in and around each problem and the whole system (system of systems).

Based on the above, this book seeks for the questions of (1) how we can address the interlinkages across SDGs or natural, human, and social systems, (2) how existing approaches or practices for SDGs can be assessed from operational points of view, where operational gaps or missing links exist, (3) how we can synthesize different approaches and other dimensions including local knowledge into a resilience approach, and (4) how we can apply the resilience approach to systemic challenges, operational gaps or missing links to change those into possibilities for transformation (see Sect. 1.2). To seek the questions, interlinkages across the goals or systems and those across relevant subsystems, risks, stakeholders, and communities will be emphasized. Based on the overarching approach, this book encourages readers to review existing approaches to SDGs from small local communities to policy practices through local to global levels from multidimensional, linkage, and boundary perspectives. The book, therefore, serves as a guide to accelerate implementation of SDGs by 2030.

1.1.3 Scope

This section provides a scope of the major terms or their relationship through this book. Since the related terms are often ubiquitous, which can be a barrier to understanding the essence or in-depth meaning of the terms or grasping the relationship, the barrier may lead to gaps in understanding the picture of relevant systems, resulting in inactions or the lack of engagements by stakeholders for challenges.

1.1.3.1 Resilience and a Resilience Approach

There are three aspects to note regarding the scope of resilience and a resilience approach. First, while massive volumes of literature for resilience focus on a narrow framing, such as the capacity to recover from disasters or the ability to withstand shocks (Alexander, 2013), resilience related to SDGs is beyond such a narrow framing and encompasses broader scope as Folke (2016) indicated that resilience is about cultivating the capacity to sustain development despite expected and surprising changes and diverse development pathways and potential thresholds between them.

If only a narrow scope of resilience is applied to the limited area, you may miss the relationships, or linkages and boundaries of nature-human-social systems, which holds a key in enabling resilience (Holling, 1973; Shimizu & Clark, 2019). Given this understanding, specific definition of resilience may include: "the capacity of a system, be it an individual, a forest, a city, or an economy, to deal with change and continue to develop" (Stockholm Resilience Center); and if you focus on more disaster risks, "the capacity of a system, community, or society potentially exposed to hazards to adapt, by resisting or changing to reach and maintain an acceptable level of functioning and structure" specified in the International Strategy for Disaster Reduction (ISDR) (2005) in the Hyogo Framework of Action (HFA) 2005–2015.

Second, because this book focuses on how to enable resilience for problem-solving oriented actions toward implementing SDGs, operational perspectives of resilience are more relevant: Operationally, resilience can be considered as the enabling capacity to create environments or systems that remain functionally intact when impacted by unexpected events, which is accomplished through recognizing situational changes and understanding "the whole system" linkages from a short-, medium- and long-term perspective (Shimizu & Clark, 2019). To enable the capacity to build resilience, critical factors include "learning to live with change and uncertainty; nurturing diversity for reorganization and renewal; combining different types of knowledge for learning; and creating opportunity for self-organization toward social-ecological sustainability" (Berkes et al., 2008).

Third, in terms of "resilience of what and to what" (Carpenter et al., 2001), which is considered a critical question to address before discussions on resilience given the broad framing, this book addresses the resilience of humans, social, and natural systems, especially their linkages, to sustainable future where human society coexist with nature through sustainable development, i.e., meeting the needs of the present without compromising the ability of future generations to meet their own needs (Brundtland & Khalid, 1987). Therefore, while SDGs include "resilience" in major goals, such as Goal 9 "build resilient infrastructure, promote inclusive and sustainable industrialization, and foster innovation and resilient cities" or Goal 11, "make cities and human settlements inclusive, safe, resilient, and sustainable," application of a resilience approach is not limited to these goals. Rather, a resilience approach directs human society's challenges across natural, human, and social systems toward problem-solving-oriented actions, specifically paying attention to linkages and boundaries among different domains or agents. Thus, the approach allows individuals, communities, or organizations to engage in challenges.

1.1.3.2 Sustainability, Sustainable Development/Sustainable Society, and Sustainable Future

In the context of SDGs or relevant fields, the term sustainability is used beyond a narrow way, such as the meaning of continuity of one agent's or sector's activities without disruption. A general sustainability theme in SDGs or relevant fields is about how we can develop or continue to improve human wellbeing and our life as a species on this planet while preserving the environment and natural resources needed by future generations (Miller, 2013). This understanding can be applied to related terms such as sustainable society or sustainable future, though different studies and scholars may express various emphases and implications. Especially since this book focuses on the transformation for systemic challenges as discussed above, those relevant terms consider the dynamics of natural, human, and social systems without compromising each other system and their interplay or synergetic process among those systems for transformation.

Moreover, the term "resilient, sustainable future, or society" tends to be used as a buzzword globally. However, at least it is important to note: "resilient and sustainable" future or society will require socioeconomic development for improved human wellbeing while preserving Earth-system resilience through nurturing human beings' coping capacities, that is, resilience, given increasing inequality within and between societies with billions left behind and rising systemic risks due to ever-increasing human pressures on the planet (TWI, 2020). Although sustainability vs. resilience is sometimes discussed, these terms are not opposite but demonstrate complementary relationships (see more in-depth discussions by Chabay in Chapter 3). Specifically, in terms of relationships between a resilience approach and sustainability, a resilience approach can play a role as an operational tool for sustainability.

1.1.3.3 "Systems", "Living Systems" and "Systemic"

As discussed above, systems/subsystems in and around SDGs are associated with natural-human–social systems. While "systems" are often considered as machines-like systems, most of those systems related to SDGs are "living systems." What does it mean by living systems? Senge et al. (2005) explain the following:

> If a part is broken, it must be repaired or replaced. This is a logical
> way of thinking about machines. However, living systems are different.
> Unlike machines, living systems, such as your body or a tree,
> create themselves. They are not mere assemblages of their parts
> but are continually growing and changing along with their elements.
> A part, in turn, was a manifestation of the whole, rather than
> just a component of it. Neither exists without the other.
> The whole exists through continually manifesting in the parts,
> and the parts exist as embodiments of the whole.

The emphasis on living systems in and around SDGs is critical since the interactions of natural-human–social systems are the underlying bases of our human society with nature. Even if we talk about policy systems or parts of social systems such as health systems or disaster management or physical systems such as bridges and dams, the systems where human and natural spheres are interrelated, are considered as living systems in that they are connected with natural phenomena, different kinds of stakeholders or communities, and their functions and relationships with other parts of the systems in different ways directly or indirectly.

Furthermore, relating to living systems, Okada (2021) provides a key in promoting our understanding of the living systems with a focus on communities. He says communities face "systemic" situations, which are driven by different layers of outside circumstances (different layers of wholeness). However, at the same time, communities can drive "systemic" dynamics from a small wholeness, through the inside and outside of the communities. As such, communities demonstrate some potential survival and self-organizing capacity, that is, "systemic capacity" (Okada, 2021).

Given SDGs cannot be implemented without communities, the systemic dynamics communities exhibit can be a driver for challenging systemic challenges. Understanding living systems is key in implementing SDGs. Also, flexible perspectives to grasp wholeness and parts from micro to macro levels across scales and their relationships are critical to address the living systems.

1.1.3.4 Disaster Risk Management and Climate Change

While SDGs do not provide a specific goal for disaster risk management, including disaster risk reduction, the theme is incorporated into the targets under different goals (see Table 1.1), which indicates that disaster risk management is one of the underlying themes across diverse goals. The list illustrates that the targets, including disaster risk management, are relevant to systemic risks driven by climate change, including social, economic, and environmental risks (see the italic in the list).

Especially, we have seen that impacts of climate change on human society become visible in the form of slow-onset disasters such as a sea-level rise, or sudden onset disasters such as hurricanes, and impacts of climate change on natures become evident in the form of the loss of biodiversity or the degradation of the environment. It is to note that all these impacts cast a shadow over our sustainable society, while climate change which impacts both humans and nature is caused by human activities (IPCC, 2021), and our sustainable future is only possible if we change trajectories through our transformative actions across different dimensions with the focus on how we can

Table 1.1 Goals and targets that incorporate specific description relating to disaster risk management

Goals/targets	Specific description relating to disaster risk management
SDG #1 (poverty) target 5	By 2030, build the resilience of the poor and those in vulnerable situations and reduce their exposure and vulnerability to *climate-related extreme events and other economic, social, and environmental shocks and disasters*
SDG#2 (hunger) target 4	By 2030, ensure sustainable food production systems and implement resilient agricultural practices that increase productivity and production, help maintain ecosystems, strengthen *capacity for adaptation to climate change, extreme weather, drought, flooding, and other disasters, and progressively improve land and soil qualities*
SDG #11 (cities) target 5	Reduce the number of deaths and the number of people affected and decrease *economic losses* relative to the global gross domestic product caused by *disasters*
SDG #11 (cities) target B SDG #13 (climate change) target 1	Increase the number of cities and human settlements adopting and implementing integrated policies and plans toward inclusion, resource efficiency, mitigation, and adaptation to *climate change, resilience to disasters, and develop and implement, in line with the Sendai Framework for Disaster Risk Reduction 2015–2030, holistic disaster risk management at all levels* strengthen resilience and adaptive capacity to *climate-related hazards and natural disasters in all countries*

co-exist with nature. Therefore, although issues of climate change or sustainability society and disaster risk management tend to be addressed separately, the linkage among climate change, sustainability, and disaster risk management or the linkage between natural, human, and social systems (or subsystems) is a critical scope for a resilience approach in this book.

1.2   From Systemic Challenges to Transformation

1.2.1   Structure of Systemic Challenges

The underlying theme is the transformation for systemic challenges, focusing on linkages across natural, human and social systems, which is the major concerns of SDGs at local through global scales. Along with the major characteristics of systemic challenges stated in the Introduction, (1) complex cause-effect structures from macro to microlevels, (2) multiple interacting components, and (3) multidimensional and cascading impacts and "deep uncertainty," specific explanation are as follows:

1.2.1.1   Complex Cause-Effect Structures from Macro to Micro Levels

The complex cause-effect structures of systemic challenges from macro to micro levels can be explained mainly from three views. First, as stated in the above, while climate change and declines in nature are drivers of systemic challenges, the changes have been caused by human actions. The recent Intergovernmental Panel on Climate Change report articulated it is unequivocal that "human influence" warmed the atmosphere, ocean, and land. Additionally, widespread and rapid changes in the atmosphere, ocean, cryosphere, and biosphere occurred. Human-induced climate change is already affecting many weather and climate extremes in every region across the globe (IPCC, 2021).

Second, other than climate change, human actions such as land and sea use, direct exploitation of organisms, pollution, and invasion of alien species have been causing the fabric of natural life and diversity to decline. Moreover, indirect human actions, such as production and consumption patterns, contributed to the decline in nature. In turn, the decline in nature poses serious risks for the quality of life of people (Díaz et al., 2019, IPBES, 2019).

Third, conversely, climate change and natural decline linked with social or economic risks cause different scales of disasters in specific local or urban areas. For example, suppose people live in mountain areas where forests are not managed well because of the lack of labor forces. In that case, extreme rainfalls driven by climate change may lead to land sliding in forests and large-scale flooding in residential areas close to forests, affecting the lives of residents in the areas. Another example is that climate change causes large-scale hurricanes and sea-level rises in urban areas. If large-scale hurricanes occur in the urban areas, the sea-level rise and disaggregated infrastructure in the urban and poor areas may largely affect the consequences of the disaster.

1.2.1.2 Multiple Interacting Components

Within the above complex structures, different natural-human–social systems/subsystems and their components (cf. natural, human, and social resources, sectors, regions, stakeholders, and communities) interact. The interaction generates nonlinear relationships among those components, resulting in dynamic evolutions at natural, human, and social levels, which is related to "living systems" as discussed in the above. This reality requires skillful coordination, consistent management, and processes by considering linkages and boundaries across risks, sectors, regions, stakeholders, and communities.

What is more, transformation (see Sect. 1.2.2), which is required for implementing SDGs, cannot be achieved without dynamic interactions of human, natural, and social systems. The dynamics are associated with livable synergies and processes. It is to note that the dynamics of systems here refer to not only articulated and visible connections of systems but also implicit and invisible relationships associated with cultural and living systems/components around boundaries of the human, natural, and place-based/community/social dimensions. Not linear but nonlinear dynamism of different systems ("living systems") and dimensions can drive the transformation.

1.2.1.3 Multidimensional and Cascading Impacts and Deep Uncertainty

The multiple interacting components within complex cause-effect structures from macro to micro levels result in multidimensional impacts on natural/ecological, human, social (including economic and developmental) dimensions in short-, mid-, and unanticipated long-term ways. In short, mid, and long terms, the impacts often cascade across different sectors and regions or from local to national (regional to global).

These multidimensional and cascading impacts lead to "deep uncertainty" in which the experts do not know or the parties to a decision cannot agree with (1) relationships among drivers and components, which shape the future, (2) function of a system, systems, and their boundaries, and (3) outcomes and their values (Lempert, 2003, Marchau et al., 2019). For example, the COVID-19, which was interlinked with urbanization, population density, and globalized travel and trade, as well as poverty and resulted in far-reaching social and economic consequences in the near future too, demonstrated that global public health risks interact with different national and local social and economic systems, which lead to the reality that the various parties to a decision do not know or cannot agree upon the system and its boundaries.

This reality brings uncertainties about how to "link" local, national, and international public policy involving different parties, including scientists, private sector, and communities (see more discussion in Chapter 4).

1.2.2 Transformation

Given the above systemic challenges, while various voices from different fields call for "changes," just calling for changes or temporary or reactive one-dot correction of a limited area will not make the transformation possible. Indeed, although the transformation tends to be used as a buzzword, little literature addresses specific approaches in terms of how the transformation is possible. As a current status, Deubelli and Mechler (2021) pointed out that reviewing the literature of transformational change and transformative approaches in the context of climate risk management and adaptation found a clear operational gap in terms of translating the transformational change into concrete transformative measures.

How can we narrow the operational gaps? While this question is related to a whole discussion on a resilience approach (see Chapter 2 for the full discussion), as a starting point to narrowing the gap, systemic challenges require "systemic" views (see the pillars of the views in Box 1), which is a baseline for a resilience approach. There are two points to note regarding the pillars of views. First, given the above complex and uncertain characteristics of systemic challenges, a transformation in the context of SDGs should entail **a process including the phases: (1) navigation of the transition, (2) building resilience of the new trajectory of development** (Olsson et al., 2004)**, and (3) systemic shifts in values and beliefs, patterns of social behavior, and multilevel governance and management regimes** (Olsson et al., 2014). Through these phases the pillars of systemic views in Box 1 need to be incorporated. Second, to operationalize the systemic views, the process of catalyzing novel pathways toward far-reaching societal transformations should involve a profound change in the scientific approaches addressing transitions to sustainability (Renn et al., 2020). The change includes **linking science, policy, and local/communities, or global and local/place-based contexts, and enabling co-knowledge production for actions**. This linkage-based process is one of the critical components of a resilience approach.

Box 1: Pillars of "Systemic" Views in the Context of SDGs

a. **Continuum view**: Systemic views in the context of SDGs require looking at natural, human and social systems in a continuum. Simply speaking, the view refers to the perspective in which nature, humans and society are not separate, but are a dynamic system of interconnected components evolving together.

b. **Focus on structures, relationships and functions**: Transformative actions for SDGs whose characteristics are systemic necessitate a system-wide reorganization across different systems/subsystems with the focus on their structures, relationships and functions, with the focus on:

   – not only direct causes but also indirect ones
   – both upstream and downstream in a flow
   – both micro and macro perspectives
   – both contexts and the whole
   – dots to lines

   The attentions to these aspects will address "linkages and boundaries."

c. **Consistent views for different stages for changes**: More specifically, toward transformation there are different stages for changes including absorption, adaptation and innovation (see the diagram below). It is through these stages that navigation of transition, building resilience, and systemic shifts can be possible for transformation. As such, it is critical to have consistent views for changes in different stages toward transformation. Importantly in order to make changes operational, it is critical to incorporate a resilience approach in each stage, including non-linear learning, co-production of knowledge and feedback loops across time and scale (see details in Chapter 2).

1.3 Toward Acceleration of SDGs

Based on the above, this book lays out pathways for addressing systemic challenges in and around SDGs or natural-human–social systems toward transformation of human society and acceleration of SDGs from the synergetic perspectives of different stakeholders. There are three points to note in drawing the picture toward acceleration.

First, although SDGs are a global framework, implementation of SDGs requires linking different stakeholders, primarily global to local linkages, which is one of the major operational gaps in existing major approaches for SDGs (see Chapter 2). **Global to local linkages** do not necessarily mean one direction from global to local or policy to community practices. It is to note that local or community practices, those of which sometimes do not use the term "SDGs," can provide critical inputs to the global community regarding how to implement SDGs, as seen from Chapters 5 and 6. The global–local linkage in this context is closely associated with the above "living systems" or dynamic interactions of human, natural, and social systems.

Second, **a resilience approach for SDGs does not necessarily refer to a new approach, but it is an approach drawn by (1) weaving different ways of thinking and approaches to face systemic challenges and (2) associating those with lessons from past experiences across regions and related cases to the systemic challenges.** The approach embraces the dynamic character of human–social-natural system interactions and elucidates the direction to identify "missing links" through different dimensions of linkages for transformation (Shimizu & Clark, 2019) (see Chapter 2).

Third, from a research viewpoint, a resilience approach is to promote **transdisciplinary research** in that the resilience approach to SDGs focuses on **(i) understanding the complexity of the problem, (ii) emphasizing the diversity of perspectives on a problem, and linking different types of knowledge through discourse, and (iii) addressing a common good** (Renn, 2021).

Through the resilience approach, readers are encouraged to identify "missing links" to for the transformation of human society toward implementing SDGs. As such, innovators in different fields, including research, educational, administrative, and business fields, can seek out the relevance of that approach to their SDG-related practices. Therefore, the book serves as a guide to how the resilience approach can accelerate the implementation of SDGs by 2030.

1.4   Chapter Introductions

Built upon the above, the book provides the following chapters under four major themes: Under the first overarching theme, *A Resilience Approach for SDGs*, following this introductory chapter, the author of this chapter provides a core part of this book: By identifying operational gaps and missing links in implementing SDGs, which include global–local linkages, Chapter 2, *Operational Gaps and A Resilience Approach for SDGs*, synthesizes a resilience approach, which can be an operational tool to narrow the operational gaps and missing links.

Thereafter, the second overarching theme, *Underlying Challenges: Global to Local Linkages,* is spotlighted. Under this theme, in Chapter 3, *Reconciling Risk, Resilience, and Sustainability: Learning from Narratives*, Chabay contributes to the book by providing the encompassing analytic views on systemic risks, resilience, and sustainability, and associates the analysis with narratives. He states that depending on normative judgments particular to the cultures and contexts in which the risks may manifest, the relationship between resilience and sustainability demonstrates the potential for synergies or tradeoffs, and important insights into the normative landscape in co-located or online communities can be gained by collecting and analyzing their narratives of imagined futures and social identities. As such, narratives can be a driver in operationalizing SDGs. Chapter 4, *Deep Uncertainty: Role of Co-knowledge Production*, addresses the issue of complex uncertainties or deep uncertainty in systemic challenges in and around SDGs, especially as risks are becoming more systemic, the systemic risks exhibit much more uncertainties, which is a common challenge in implementing SDGs. With the focus on co-knowledge production that is a leverage point of global–local linkages and a resilience approach, Chapter 4 associates the management of deep uncertainty with a resilience approach, and it demonstrates how the resilience approach can be incorporated in a specific community program related to SDGs and how the approach or co-knowledge production is applicable to facing the COVID-19.

For the following chapters, from Chapter 5 through 8, authors with different disciplinary backgrounds and experiences contribute to this book. The third overarching theme is *Practices Nurtured through Locals and Communities*. Under this theme, in Chapter 5, Okada focuses on community resilience from the viewpoint of community's coping capacity challenged for surviving and regrowing toward a sustainable future, with conceptual and methodological frameworks and field-based evidence. In Chapter 6, Nakamura provides two field explorative case studies, local stakeholder dialog on volcanic disaster management, and citizen dialog with experts on radioactive waste. Based on these cases, the chapter proposes methods to develop a culture of dialog that nurtures resilience. Both Okada and Nakamura focus on creating an environment by linking scattered resources and looking at details and the whole. Especially these two chapters tell us that global to local linkages should *not* necessarily be from the top to down or *not* necessarily from one direction from global or national or policy to locals.

Under the fourth overarching theme, *Progress, Possibilities and Challenges in Education*, in Chapter 7, *Local Community-Based Education for Sustainable Development (ESD) During the COVID-19 Pandemic in the Asia and Pacific Region*, Noguchi illustrates how the adaptive approach (which is relevant to a resilience approach) has been developed during the COVID-19 drawing the experience of the Regional Centre for Expertise on ESD (RCEs) in the Asia and Pacific region by United Nations University. The chapter discusses the factors that promote or hinder the activities, including the issue of "digitalization." Chapter 8, *Shaping Sustainability Priorities for Higher Education Institutions*, by Vincent, explores the framing of sustainability priorities in relation to the complex functions of higher education institutions by analyzing institution-level SDGs-aligned strategies published by universities in different parts of the world. Based on this analysis, this chapter concludes that developing resilience-based approaches is necessary and urgent.

As a conclusion, Chapter 9 by the author of this chapter, reflecting all chapters, seeks to the concluding question: How can a resilience approach address SDGs? Through transdisciplinary research viewpoints, the final chapter spotlights different stakeholders, and most importantly, this book's sub-theme, global to local linkages, local communities. The book demonstrates how a resilience approach can be incorporated in their related efforts to SDGs by articulating what is missing in existing efforts.

References

Alexander, D. E. (2013). Resilience and disaster risk reduction: An etymological journey. *Natural Hazards and Earth System Sciences, 13*(11), 2707–2716.

Berkes, F., Colding, J., & Folke, C. (Eds.). (2008). *Navigating social-ecological systems: Building resilience for complexity and change.* Cambridge University Press.

Brundtland, G. H., & Khalid, M. (1987). *Our common future.* Oxford University Press.

Carpenter, S., Walker, B., Anderies, J. M., & Abel, N. (2001). From metaphor to measurement: Resilience of what to what? *Ecosystems, 4*(8), 765–781.

Deubelli, T. M., & Mechler, R. (2021). Perspectives on transformational change in climate risk management and adaptation. *Environmental Research Letters, 16*(5), 053002.

Díaz, S., Settele, J., Brondízio, E. S., Ngo, H. T., Agard, J., Arneth, A., Balvanera, P., Brauman, K. A., Butchart, S. H. M., Chan, K. M. A., Garibaldi, L. A., Ichii, K., Liu, J., Subramanian, S. M., Midgley, G. F., Miloslavich, P., Molnár, Z., Obura, D., Pfaff, A., Polasky, S., Purvis, A., Razzaque, J., Reyers, B., Chowdhury, R. R., Shin, Y.-J., Visseren-Hamakers, I., Willis, K. J., & Zayas, C. N. (2019). Pervasive human-driven decline of life on earth points to the need for transformative change. *Science, 366*(6471), eaax3100

Folke, C. (2016). Resilience (republished). *Ecology and Society, 21*(4), 1–30.

Grainger-Brown, J., & Malekpour, S. (2019). Implementing the sustainable development goals: A review of strategic tools and frameworks available to organisations. *Sustainability, 11*, 1381.

Holling, C. S. (1973). Resilience and stability of ecological systems. *Annual Review of Ecology and Systematics, 4*, 1–23.

Intergovernmental Science-Policy Platform on Biodiversity and Ecosystem Services (IPBES) Secretariat. (2019). *Summary for policymakers of the global assessment report of the Intergovernmental Science-Policy Platform on Biodiversity and Ecosystem Services.*

IPCC. (2021). *Climate change 2021: Summery for policymakers.* https://www.ipcc.ch/report/ar6/wg1/. Accessed on 12 October 2021.

Lempert, R. J. (2003). *Shaping the next one hundred years: New methods for quantitative, long-term policy analysis.* RAND.

Marchau, V. A., Walker, W. E., Bloemen, P. J., & Popper, S. W. (2019). *Decision making under deep uncertainty: From theory to practice.* Springer Nature.

Miller, T. R. (2013). Constructing sustainability science: Emerging perspectives and research trajectories. *Sustainability Science, 8*(2), 279–293.

Okada, N. (2021, November 26). *Another challenge: Systemic thinking and design for community-based/humans-focused disaster risk governance.* Invited speech a DPRI SOGO-Bosai seminar 50th Session.

Olsson, P., Folke, C., & Hahn, T. (2004). Social-ecological transformation for ecosystem management: The development of adaptive co-management of a wetland landscape in southern Sweden. *Ecology and Society, 9*(4), 2.

Olsson, P., Galaz, V., & Boonstra, W. J. (2014). Sustainability transformations: A resilience perspective. *Ecology and Society, 19*(4), 1.

Renn, O. (2021, April). Transdisciplinarity: Synthesis towards a modular approach. *Futures, 130*, 102744.

Renn, O., Chabay, I., van der Leeuw, S., & Droy, S. (2020). Beyond the indicators: Improving science, scholarship, policy and practice to meet the complex challenges of sustainability. *Sustainability, 12*(2), 578.

Senge, P. M., Scharmer, C. O., Jaworski, J., & Flowers, B. S. (2005). *Presence: An exploration of profound change in people, organizations, and society.* Currency.

Shimizu, M., & Clark, A. L. (2019). *Nexus of resilience and public policy in a modern risk society.* Springer.

Trump, B. D., Keenan, J. M., & Linkov, I. (2021). Multi-disciplinary perspectives on systemic risk and resilience in the time of COVID-19. In *COVID-19: Systemic risk and resilience* (pp. 1–9). Springer.

TWI2050—The World in 2050. (2020). *Innovations for sustainability: Pathways to an efficient and post-pandemic future.* Report prepared by The World in 2050 initiative. International Institute for Applied Systems Analysis (IIASA), Laxenburg, Austria. www.twi2050.org

Mika Shimizu is an Associate Professor in Graduate School of Advanced Integrated Studies in Human Survivability, Kyoto University. Her long years' experiences as a policy researcher in East-West Center in Washington DC and Honolulu, Hawaii in the United States greatly contributed to publishing this book. She holds an M.A. from American University and a Ph.D. in International Public Policy from Osaka University (2006). She has been extensively involved in interdisciplinary and transdisciplinary research projects related to disasters/infectious diseases, sustainability, and climate change issues with the focus on resilience. Her major publications include *Nexus of Resilience and Public Policy in a Modern Risk Society* (Co-Author: Allen Clark, Springer, 2019).

Chapter 2
Operational Gaps and a Resilience Approach for SDGs

Mika Shimizu

Abstract This chapter assesses the current state of implementation of Sustainable Development Goals (SDGs) and the existing approaches from an operational perspective to derive the operational gaps. Further, it provides a framework of a resilience approach for SDGs to bridge these operational gaps. The aim of a resilience approach for SDGs is to address systemic challenges (see Chapter 1) in and around SDGs to develop problem-solving oriented actions for transformation by operationalizing resilience, i.e., nurturing resilience in different layers of societies in harmony with nature. With regard to SDGs, nature, humans, societies, and communities have originally capacities to change, adapt, or recover; however, those capacities can be weakened or lost if they are impacted beyond their threshold by stresses, shocks, or adversities interrelated with systemic challenges surrounding SDGs. To avoid this situation, the resilience approach, which was developed to operate beyond disciplines and different kinds of knowledge including local and indigenous knowledge will contribute to formulating problem-solving oriented actions.

2.1 Introduction

As discussed in Chapter 1, the aim of a resilience approach for SDGs is to address systemic challenges (see Chapter 1) in and around SDGs to develop problem-solving oriented actions for transformation by operationalizing resilience, i.e., nurturing resilience in different layers of societies in harmony with nature. The application of a resilience approach is *not* limited to specific goals that include the term "resilience" or "resilient" in their name—for instance, Goal 9 "build resilient infrastructure, promote inclusive and sustainable industrialization, and foster innovation and resilient cities," or Goal 11 "make cities and human settlements inclusive, safe, resilient, and sustainable." Rather, a resilience approach can be applied to address SDG-related systemic challenges involving linkages and boundaries among different domains or agents,

M. Shimizu (✉)
Graduate School of Advanced Integrated Studies in Human Survivability, Kyoto University, Kyoto, Japan

especially by considering natural, human, and social systems in a continuum across time and scale regardless of the specific number of goals.

The contributions of this chapter are three-fold. First, it delineates the operational gaps in terms of the current state of implementation of SDGs—in other words, the manner in which the SDGs have been addressed—by reviewing existing studies and examining the characteristics of existing practices from an operational perspective. In this context, "operational" refers to "problem-solving oriented," taking into account the systemic challenges related to SDGs. Second, relevant approaches to address linkages between SDGs are assessed from an operational perspective; these approaches have been emerging recently and are key to implementing the SDGs. Third, a resilience approach for SDGs—the framework underlying a resilience approach for implementing or accelerating the implementation of SDGs—is provided, which, in turn, will address the operational gaps. The details of the resilience approach include its premises, foundations, operational pillars, and application to practice. The resilience approach provided here in the context of SDGs, is a synthesis of the modes of resilience-related thinking and practices across disciplines, including systems approach, and is shaped by the diverse experiences of human societies as well as the lessons learnt across nations/regions and communities.

2.2   Operational Gaps in Implementing SDGs

To articulate the current state of implementation of SDGs, the following sections present major findings of existing studies related to implementation of SDGs, and draw the characteristics of existing approaches that impede the progress of implementation of SDGs using an example of linkage challenges, climate change and forests. On this basis, the operational gaps in implementing SDGs will be identified.

2.2.1   Findings in Existing Studies Related to Implementation of SDGs

Major findings of existing studies directly or indirectly related to the current state of implementation of SDGs are compiled in Table 2.1, which presents the research themes and findings for respective studies.

Overall, Table 2.1 demonstrates that (1) limited research and analyses/assessments focus on linkages among SDGs (see A, B and D), and more importantly that, (2) limited tools are available to translate words into specific actions and address the linkages among SDGs (see C, E). Reflecting on the identified limitations, the implementation of SDGs is still in its nascency, specifically in terms of undertaking problem-solving oriented actions that promote implementation of SDGs. This is especially reflected in the assessments in the recent voluntary national review reports (see

Table 2.1 Major findings in existing studies directly or indirectly related to practices involving SDG implementation

Studies (author, year)	Research theme	Findings
(A) Weitz et al. (2018)	SDG interactions with respect to its systemic and contextual characteristics	Research on interaction among the SDGs has been very limited. Studies typically focus on a specific goal area and explore its links with other SDGs. However, a systemic and contextual analysis of SDG interactions indicated that **considering the systemic impacts could influence decisions regarding which efforts should be prioritized to enhance the effectiveness of implementation strategies**
(B) C. Allen et al. (2018)	Reviews of academic and expert literature as well as country-level experiences in implementing the SDGs	While progress has been made in some initial planning stages, key gaps remain with regard to assessing the interlinkages, trade-offs, and synergies between targets. **Gaps were also identified in the adoption of systems thinking and integrated analytical approaches and models**

(continued)

Table 2.1 (continued)

Studies (author, year)	Research theme	Findings
(C) J. Grainger-Brown and S. Malekpour (2019)	Tools available to organizations for implementing SDGs	Most of the tools are only applicable to "mapping" and "reporting" activities, which are activities that occur after strategies have been developed and even implemented; The study could not identify tools that substantially engaged with **actual strategy development**—the stage that **can shape transformative change**
(D) A. De Oliveira and S. Kindornay (2021)	Assessment of the voluntary national review (VNR) reports submitted to the United Nations High-level Political Forum on Sustainable Development in 2020	While improved reporting on best practices and learning from peers is a positive gain, the reports indicated the limited planning for implementing SDGs and addressing encountered difficulties. Fewer countries (compared with previous assessments) reported conducting baseline and **gap assessments** and provided information on data availability. Overall, information on national, regional, and global **follow-up and review processes** suffered from backsliding. Most VNR reports lack reference to accountability mechanisms at the national level
(E) T. M. Deubelli and R. Mechler (2021)	Climate risk management and adaptation	Reviewing literature in the context of climate risk management and adaptation; the study found a clear operational gap in terms of **translating transformational change into concrete transformative measures**

D) submitted by 45 countries to the United Nations High-level Political Forum on Sustainable Development (Other countries do not even submit the report.). These reports highlight (a) the lack of planning for implementing and addressing encountered difficulties, (b) the lack of assessments and relevant information and data, and (c) the lack of follow-up and review processes (see D).

2.2.2 Characteristics of Existing Practices Derived from an Operational Perspective

The implementation of SDGs is assessed from operational points of view, i.e., from the perspectives of problem-solving oriented actions toward implementing SDGs. Two major dimensions have been emphasized given the systemic challenges in and around SDGs (see Chapter 1), namely, **(1) addressing the systemic challenges in and around SDGs, and (2) incorporating systemic views into practices to address the systemic challenges** in specific ways.

As discussed in Chapter 1, the pillars of the systemic views in the context of SDGs include the following **basic but critical aspects: (a) looking at (considering) natural, human, and social systems in a continuum, (b) a system-wide reorganization across different systems/subsystems with a focus on their structures, relationships, and functions by addressing linkages and boundaries in systems/subsystems, and (c) consistent views for different stages of transformation. To operationalize these systemic views, a systems approach is essential**. The systems approach emphasizes that a singular focus on components does not provide a whole picture of complex problems and takes into account a multidimensional means of considering the various interacting components (Lim et al., 2018; See more discussions in Sect. 2.4).

From the operational perspective based on the above baseline, the following three characteristics can be identified in the existing practices based on existing relevant international reports and research papers.

2.2.2.1 Characteristic 1: Awareness of Linkages Among SDGs Not Linked with Operational Actions

The scientific community has underscored the necessity of addressing interlinkages among SDGs based on systems thinking (Allen et al., 2018). As a major effort to inform the high-level policy community, the International Council for Science (2017) developed a framework which identifies categories of causal and functional relations underlying progress or achievement of goals and targets and provided specific examples to address interlinkages. In this manner, the scientific community's knowledge has been translated into international policy communities at least conceptually. However, this knowledge is not necessarily linked with operational actions. Even for a major challenge of SDGs, namely climate change, the linkage-based actions are still at a nascent stage. This is reflected in the challenge of climate change and forests as follows.

The recognition of the linkages between climate change and forest loss has increased globally since the New York Declaration on Forests (NYDF) was launched in 2014; they is a partnership of governments, multinational companies, civil society, and indigenous communities, who strive to halve deforestation by 2020 and to end it by 2030. Moreover, the 2015 Paris Agreement on Climate Change focused mainly on forests in Article 5. However, almost seven years later, the recent NYDF assessment revealed forest loss through deforestation, and that forest degradation has continued mostly unabated (NYDF Assessment Partners, 2021). To address this problem, the 26th UN Climate Change Conference of the Parties (COP26) in Glasgow in 2021 adopted the Glasgow Leaders' Declaration on Forests and Land Use to "emphasize the critical and interdependent roles of forests of all types, biodiversity and sustainable land use in enabling the world to meet its sustainable development goals." This declaration is a step forward as more than 140 countries endorsed it (as of November 2021) (https://ukcop26.org/glasgow-leaders-declaration-on-forests-and-land-use/), including the Russian Federation, Brazil, Canada, the US, and China, which account for 54% of the world's forests (https://www.fao.org/forest-resources-assessment/2020/en/), with more than $20 billion of public and private funds (UK Government, 2021). However, operational actions based on the above assessment results by NYDF have only begun.

2.2.2.2 Characteristic 2: Limited Attention to Different Systems/Subsystems Beyond Goal-to-Goal Linkages Among SDGs, Especially the Macro–Micro Level Linkages

The scientific community identified "systems approach" as an instrument to work on different interactions. The International Council for Science (2017) articulated that it is critical to work on SDGs beyond binary interactions "recognising that interactions can be far more complex, multidimensional and dynamic with feedbacks and unforeseen consequences," "further work on interactions could usefully apply a systems-approach," and "a systems approach can be taken at various organisational levels depending on the goals and targets and the spillover to other goals and targets." (p. 224).

However, the recognition is little operationalized yet across scale. Although major international or policy reports touch on a systems approach in addressing SDGs, a systems approach tends to be applied at the national and macro levels to map the interrelations among the goals. It has not been incorporated into practice yet to address issues beyond goal-to-goal interactions and especially, to go beyond national levels. Among the limited tools that go beyond an analysis of goal-to-goal interactions, the iSDG tool developed by the Millennium Institute is an integrated macro-economic model built upon the systems approach for SDG planning; it provides explorative scenarios and simultaneous simulation of policies developed in different sectors to analyze their synergies. An analysis of iSDG in Australia revealed how different approaches to development alter Australia's development trajectory and SDG performance (Allen et al., 2019). However, the method has a major limitation as it only targets the national-level and macro policies.

Operationally, the systems approach can be applied to broader dimensions including the overarching perspectives of human and social systems, and the macro–micro linkages such as global-national-local or policy-community linkages, which are critical "boundary" areas in implementing SDGs.

2.2.2.3   Characteristic 3: Limited Attention to Living Systems for Transformation

With regard to Characteristic 2, even some existing guidelines and tools that pay attentions to linkages in SDGs do not necessarily take into account "looking at natural, human and social systems in a continuum," or synergies of natural-human-social systems, more specifically "living systems." As discussed in Chapter 1, systems/subsystems associated with SDGs are regarded as not machine-like systems but "living systems" that create themselves and are continually growing and changing through interactions within the natural-human-social systems (Senge et al., 2005). Furthermore, communities constitute the core of living systems and are a driver of transformation. Although communities face "systemic" situations that are driven by different layers of external circumstances, but, at the same time, communities can drive transformation, even with a small step, through systemic interactions within and outside the communities (Okada, 2021). This aspect of living systems has received limited attention in discussions in the context of implementing SDGs.

2.2.3   Operational Gaps

Based on the above analysis, in a nutshell, operational gaps exist in (1) actions based on awareness of linkages of SDGs, (2) actions related to macro–micro level linkages, and (3) actions based on attention to living systems for transformation. Although some research and analyses/assessments focus on the linkages among SDGs, more in-depth perspectives regarding the linkages, based on an understanding of systemic challenges and relevant actions, are essential for implementing SDGs.

However, it is notable that (a) emerging approaches are attempting to improve the existing pathways in addressing SDGs by incorporating systemic view(s) in the approaches, and (b) macro-level trends for SDGs do *not* necessarily reflect relevant efforts at a micro level or at local/community levels. Actually, there are relevant small-scale good practices which do not necessarily adopt the term SDGs but have been nurtured locally or in communities. Section 2.3 discusses the emerging trends and practices related to (a); with regard to (b), relevant case studies are provided in Chapters 5 and 6.

2.3   Emerging Approaches to Address Linkages

With regard to addressing the linkages associated with SDGs, several approaches have been introduced or incorporated, some of which overlap in terms of scope/focus. Table 2.2 highlights, from an operational perspective, the major approaches that were introduced or adopted for addressing SDGs or relevant issues, namely, (1) co-benefits approach, (2) resilience dividend, (3) nature-based solutions (NbS), and (4) nexus approach. It reports the scope/focus and operations of each approach, and where the operational gaps exist, based on the discussion of systemic challenges that are interlinked with SDGs, here and in Chapter 1.

In sum, Table 2.2 illustrates that resilience dividend and NbS became policy instruments to link issues that used to be addressed separately before and to embody co-benefits or multi-benefits; however, the operational perspective should be explored further. The NbS approach was originally focused on climate change, but of late, it has broadened its scope to cover disaster risk reduction (DRR).

Notably, mainstream policy at the global field, as seen in Table 2.2, began to pave the way for addressing climate change and DRR simultaneously, even though many challenges and uncertainties exist from an operational perspective. The European Union (EU), which is a promoter of NbS and originally promoted this approach in climate change, expect that adopting NbS both for DRR and climate change adaptation contribute to the paradigm shift and are/should be an inherent part of future sustainable development.

In terms of how NbS can be incorporated into DRR, the UN Office of Disaster Risk Reduction (UNDRR) (2021) explains that the term Eco-DRR has been used to enhance effectiveness of DRR; combining these "green" approaches as NbS and investing on engineered structures and solutions not only achieves DRR but also responds to climate change, in addition to providing other benefits, such as the preservation of natural resources. Furthermore, the UNDRR (2021) pays attention to systemic risks in adopting NbS by expressing that a more systemic approach and understanding is required and NbS can play a role in addressing systemic risks because they involve working with the socio-ecological system as a whole.

However, the abovementioned progress is still in its early stages from an operational perspective. The operational gaps include **how to co-learn and co-design with stakeholders (Hillgren et al., 2011), how to link nature and human society, including communities and different regions, or how to link macro and micro actions.** These operational issues **cannot be solved using the top-down or conventional approach**. Notably, the move to integrating measures associated with both climate change and deforestation/biodiversity issues is still nascent at the global level as discussed above.

Table 2.2 Emerging approaches to address linkages

	Scope/focus	Operations (application to real problems)	Merits and operational gaps
(1) Co-benefits approach	• It focuses on several benefits in addition to the benefit produced by a business-as-usual scenario • It has been discussed in different areas	It is mostly adopted in the literature on climate change (Fung & Helgeson, 2017). However, even in the climate change literature, there is little trans-disciplinary work that considers the politics and institutional aspects of co-benefits other than the economic aspects (Mayrhofer & Gupta, 2016)	• It basically tells us how one problem is related to another /others, but does not provide the necessary operational solutions to enable co-benefits • However, the approaches specified below in (2) and (3) were derived from the co-benefits Approach, and these new approaches enhance and extend this approach
(2) Resilience dividend	• It focuses on the co-benefits of resilience investments in the absence of a disruptive incident (Fung & Helgeson, 2017) • It has been discussed mainly in the disaster management and development fields	As a recent development, the resilience dividend approach has been incorporated into policy strategies on disaster management and development, such as the "triple dividend of resilience" by the World Bank, which focuses on (1) avoided or reduced losses, in the event of a disruptive event; (2) increased economic resilience, from reducing disaster risk; and (3) co-benefits for development (Tanner et al., 2015)	• The resilience dividend approach (derived from the co-benefits approach) help communication across fields and bridges the gap between research and practical on-the-ground applications and planning (Helgeson & O'Fallon, 2021). However, quantifying the resilience dividend remains an issue (Fung & Helgeson, 2017) • Specifically, from an operational perspective, issues may include how to apply this approach at micro levels, such as in communities, and other areas of SDGs, and how to reflect place-based contexts and intangible or non-economic assets

(continued)

Table 2.2 (continued)

	Scope/focus	Operations (application to real problems)	Merits and operational gaps
(3) Nature-based solutions (NbS)	• Derived from the co-benefits approach, NbS aims at multi-benefits and involves "actions to protect, sustainably manage and restore natural or modified ecosystems, that address societal challenges effectively and adaptively, simultaneously providing human well-being and biodiversity benefits" (IUCN, 2016) • It has increasingly been used as an umbrella term encompassing different ecosystem-based approaches, such as green infrastructure, natural capital, and ecosystem-based disaster risk reduction (Eco-DRR) (Wild et al., 2020)	The European Union (EU) has promoted NbS as a key policy instrument in linking actions for climate change mitigation/adaptation and biodiversity • The EU recently broadened its applications to disaster risk reduction by presenting "ecosystem-based adaptation and ecosystem-based disaster risk reduction as ways of enhancing synergy between climate change adaptation and disaster risk reduction" (Zhongming et al., 2021) • In parallel with the EU policy direction, the UN Office of Disaster Risk Reduction (UNDRR) adopted NbS in policy instruments to implement the Sendai Framework for Disaster Risk Reduction 2015–2030 and provided a guideline called "Nature-based Solution for Disaster Risk Reduction" in 2021	• NbS plays a major role in promoting linkage perspectives in connecting nature and actions both for climate change and disaster risk management at the global level, including policy and business • Through NbS, measures for climate change adaptation and disaster risks are getting integrated; these used to be addressed separately • However, there is a lack of clarity on how to make NbS operational, including how to co-learn and co-design with stakeholders in different contexts. Also, there has been minimal discussion on how to link nature and human society, including communities and different regions, or how to link macro and micro actions
(4) Nexus approach	• This approach gained prominence during the Bonn 2011 conference on "Water, Energy and Food Security Nexus," which emphasized the importance of improving management and governance across sectors and scale of operations, reducing trade-offs, and building synergies (Hoff, 2011)	This approach is not well defined and the term "nexus" tends to be used not at operational levels, but more at thematic levels	• This approach is similar to the co-benefit approach, but in addition to the co-benefits, this approach also focuses on reducing the risk that contributions to one SDG undermine progress on another ("trade-offs" or "SDGs washing") • From an operational perspective, few solutions are provided

2.4 Framework of a Resilience Approach for SDGs

2.4.1 Overview

A resilience approach will narrow the aforementioned operational gaps (Sect. 2.2.3). As mentioned in Chapter 1, the resilience approach does not present a single uniform solution to fix problems, but is a synthesis of different kinds of knowledge and practices, including systems approach, systems thinking, design thinking, resilience thinking, and other relevant practices (Shimizu & Clark, 2019), to navigate the systemic challenges or the dynamics of natural-human-social systems. Thus, in the context of SDGs, the six premises underlying the resilience approach defined in this chapter are as follows:

Premise 1: The resilience approach specified here has been shaped (based on design-thinking) by interweaving different ways of thinking and approaches aimed at enabling resilience, and through interactive process between these diverse ways of thinking, approaches, and practices (including lessons learned and field and case studies, both at the micro and macro levels) (see Fig. 2.1).

Premise 2: Among the diverse existing ways of thinking and approaches, the systems approach is an underlying core conceptual tool for challenges surrounding SDGs, which are systemically interlinked. Although there are different conceptualizations of the systems approach, the followings constitute the core basis upon which the resilience approach in the context of SDGs is built (Jackson, 2010; Shimizu & Clark, 2019):

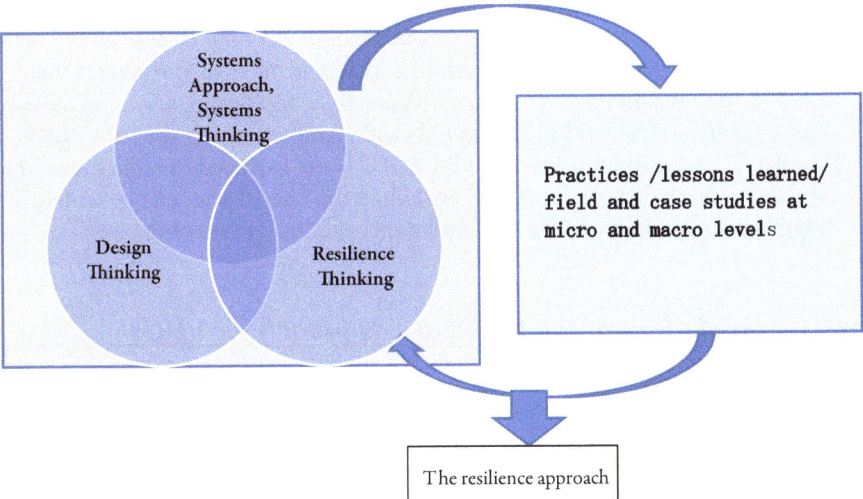

Fig. 2.1 Shaping the resilience approach

- Each system and its subsystems related to the challenges surrounding SDGs need to be identified and need to function independently, as it is essential that there is a centralized coordination of the overall system.
- Within the overall system, (1) interactions among systems and subsystems, (2) environment surrounding the systems and subsystems, and (3) boundaries between systems and/or among subsystems need to be identified and analyzed.
- Emergent characteristics of the overall system based on (1), (2), and (3) need to be identified.
- The overall system must be both continually assessed and modified/renewed to ensure functionality.

The systems approach framework enables "**looking at the details and the whole in a continuum**." Through this process, **resilience can be accomplished by recognizing situational changes and understanding "whole system" linkages from a short, medium, and long-term perspective**; this has been identified recently through different complex challenges, including disasters and climate change-related challenges (Shimizu & Clark, 2019).

Premise 3: Since the linkages and boundaries in diverse contexts across time and scales are key to operationalizing resilience, the resilience approach built upon the systems approach specifically highlight the (a) linkages/boundaries across natural-human-social systems, (b) global to local linkages/boundaries, (c) linkages/boundaries among stakeholders, including among science and policy community and local community, (d) linkages/boundaries among different kinds of knowledge (from scientific, academic, and practical knowledge through indigenous knowledge) and actions, (e) interrelations among climate change and disasters or resilience and sustainability, and (f) linkages/boundaries among various disciplines including natural, social, and human systems, sectors and actors.

Premise 4: Specifically, linkages/boundaries entail not only issues/risks/system linkages or boundaries, but also diverse components that enable resilience, including natural, human, and social/financial/technological resources, stakeholders, places/communities, generations and processes across time and scales.

Premise 5: The resilience approach for SDGs is not intended to replace existing approaches and practices for SDGs, including the ones discussed above, but for strengthening existing ones or addressing missing links in existing ones.

2.4.2   Foundations of the Resilience Approach for SDGs

The following four foundations of the resilience approach for SDGs specify the structural framework of the resilience approach based on Sect. 2.4.1. This provides the basis for the functions and operations of the resilience approach, which can be applied at any level, ranging from policy at the community level, or from macro through micro levels.

1. **Relationships and interrelated capacities**: Holling (1973) paved the way for resilience thinking and differentiated between stability and resilience; stability represents the ability of a system to return to an equilibrium state after a temporary disturbance, while resilience is a measure of the persistence of systems and of their ability to absorb change and disturbance and still maintain the same relationships between populations or state variables. Thus, the essence of resilience is in **"relationships" and furthermore, interrelated capacities of different systems**: Specifically, humans, ecology/nature, and communities/societies commonly have **capacities to change, adapt, recover, and transform**, but their capacities could be harmed or broken by persistent stresses or shocks or adversaries; **these capacities can be protected, nurtured, and strengthened only if they maintain interdependent "relationships" within the nature-human-social systems** (These relationships are illustrated using arrows in Fig. 2.2).

2. **"Looking at natural, human, and social systems in a continuum"**: "Relationships" is interlinked with "continuum" **based on the systems approach**: Paying attentions to "relationships" and associating those with Premise 2 will lead to (a) "looking at natural, human, and social systems in a continuum," that is, not addressing each system separately, but considering the relationships and the whole picture through an all-encompassing perspectives. This will lead to (b) the system-wide reorganization across different systems/subsystems, and (c) consistent views for different stages for transformation (see Chapter 1) (The whole picture is depicted in Fig. 2.2).

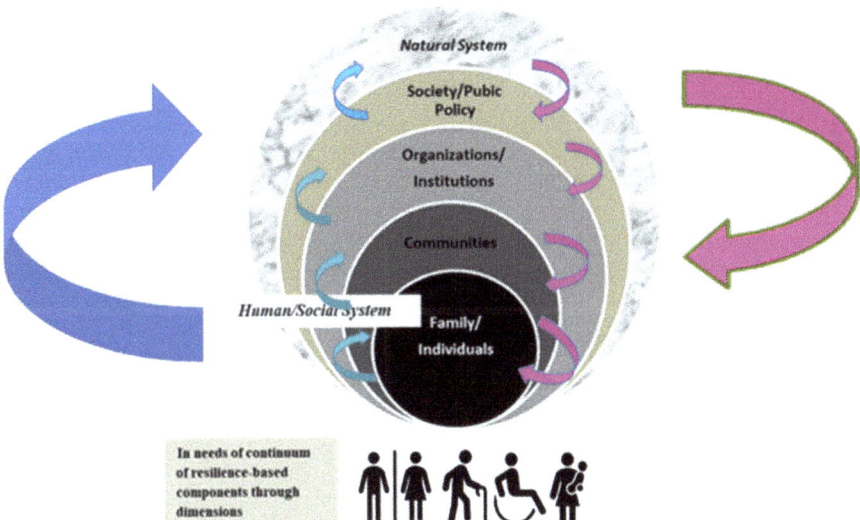

Fig. 2.2 Relationships and interrelated capacities in "looking at natural, human, and social systems in a continuum" (Shimizu & Clark, 2019, revised)

3. **Dynamism or Diversity**: The resilience approach in the context of SDGs is about **preparing for opportunities or creating conditions that allow for dynamism in the natural-human-social systems to emerge through the interactions of individuals, communities, and societies, and through their interplay within the biosphere and across scales. This interplay generates relational changes in systems/subsystems that are directed toward adaptation and transformation** (Folke, 2016) **by paying special attention to vulnerable populations and communities. To make this possible, it is important to incorporate resilience-enabling components** (for details, see Sects. 2.4.3 and 2.4.4, which specifies the components that enable operational resilience) **in different layers of systems/subsystems in a continuum** (see Fig. 2.2). Further, it is critical to note that the natural system underpins all dimensions of human health and contributes to non-material aspects of quality of life, which includes inspiration and learning, physical and psychological experiences, and supporting identities that are central to quality of life and cultural integrity, even if their aggregated value is difficult to quantify. The diversity of nature maintains humanity's ability to choose alternatives in the face of an uncertain future (IPBES, 2019).

4. **Dynamic capacity to devise**: Based on the dynamism or diversity perspective, Holling (1973) suggested that the resilience framework is not for a precise capacity to predict the future, but for **a qualitative dynamic capacity to devise systems** that can absorb and accommodate future events, **whatever unexpected form they may take**. Devising systems can be for different forms including not only absorption but also adaptation and innovation toward transformation (see Fig. 2.3).

5. **Multiple pathways**: As the resilience approach embraces dynamic characters of natural-human-social systems and human-ecosystem interactions, the approach entails multiple potential pathways within them. The resilience approach provides

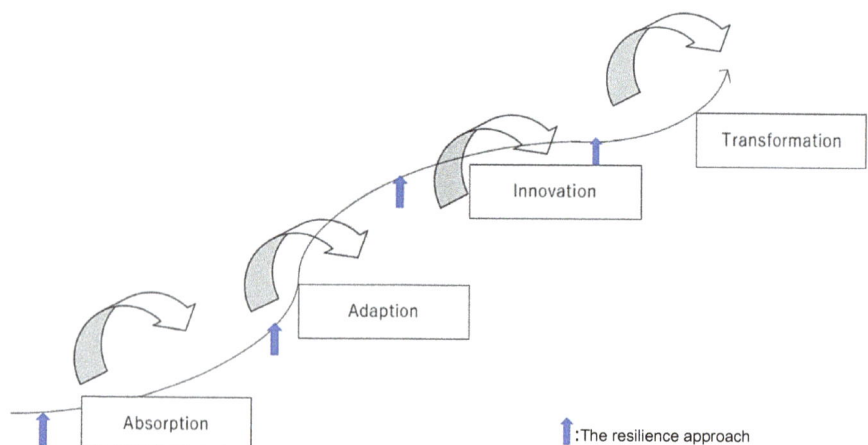

Fig. 2.3 Dynamic capacity to devise (Shimizu, 2015, revised)

courses in terms of how individuals, communities, and societies' positive response to change can be strengthened and supported. Regarding the multiple pathways, Holling (1973) also underlined a management approach based on resilience, which would emphasize the need to **keep options open and the need to maintain heterogeneity**.

6. **Modular, non-linear, and consistent process**: The resilience approach **can be applied to different stages of changes including absorption, adaptation, and innovation to achieve transformation in different ways depending on contexts; at the same time, the approach emphasizes a modular and consistent process** (see Fig. 2.3). In this process, **structuring co-learning processes in teams,** including **observation, prototyping, and testing or soliciting feedback** based on design thinking (Plattner, 2010 and Hillgren et al., 2011) **through "triple-loop learning"** is critical.

Unlike the incremental improvement of action strategies without questioning the underlying assumptions (single-loop learning) and a revisiting of assumptions (double-loop learning), the triple-loop learning **reconsiders underlying structures and paradigms that go beyond the established framework** (Pahl-Wostl, 2009). As such, **the non-linear approach should be adopted** by encouraging policy planners and/or civil society partners to imagine various futures and contemplate their uncertainty (Shimizu & Clark, 2019).

2.4.3 Operational Pillars

Specific or detailed operational pillars for incorporating the resilience approach in (re)planning, (re)designing or (re)structuring SDG-related projects/programs are discussed below. The pillars can also be applied at any level, ranging from the macro level including different policies to the micro level including community, organizational, and project/program levels, by paying attention to the macro-micro linkages

Operational Pillar 1: "Looking at natural, human, and social systems in a continuum" views: **more specifically, "looking at nature/ecology, different kinds of humans and communities/societies in a continuum"** can contribute to ensuring that actions are *not* **at the expense of certain aspects of the SDGs, i.e.,** *not* **causing trade-offs.** This is specifically because the views are built upon the understanding **that the capacities of natural, human, and social systems to change, adapt, recover, and transform can be developed only through their interrelationships.**

Specifically, addressing "the relationship" needs to take into account both the whole picture (natural, human, and social systems) and its parts from different dimensions, including **(1) linkage, (2) process, (3) temporal (time), and (4) scale perspectives** (see Fig. 2.4). Each lens is best understood in relation to one another in an integrated fashion, while individual components may already have

Fig. 2.4 Critical dimensions
in enabling resilience
(Shimizu, 2015)

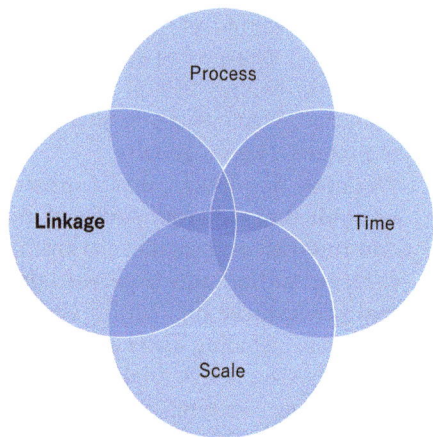

been utilized in different literatures. The critical point here is to look at the whole
in a balanced manner, using the four critical lenses that provide the necessary
components to review the interaction of these components (Shimizu & Clark,
2019).

Operational Pillar 2: The relevant actions for SDGs should take into
account different **contexts** surrounding nature, humans, and societies, **including
geographical/social/economic/cultural/ historical contexts, with a focus on
vulnerable populations, including the aged, disabled, poor, and other vulnerable people or communities.**

Operational Pillar 3: **Co-learning** processes involving **different stakeholders
with triple-loop learning** (see Sect. 2.4.2 (6)). Furthermore, building/enabling a
co-knowledge or collaborative knowledge production scheme or system based
on the co-learning scheme, which can bridge across the boundaries of disciplines
and different kinds of knowledge, is a **leverage point** in **nurturing or operationalizing resilience** (Shimizu & Clark, 2019)—**in other words, in addressing
systemic challenges in and around SDGs towards problem-solving oriented
actions for transformation** (see Sect. 2.4.4).

Operational Pillar 4: To operationalize the resilience approach, and especially, to
enable co-knowledge or collaborative knowledge production, **"enabling environments"**—understood as a **set of interrelated conditions, such as organizational,
social, environmental, and cultural conditions—are important** as the **environments or the interrelated conditions impact the capacity of stakeholders to
engage in processes related to SDGs.**

2.4.4 Co-knowledge Production to Link Macro and Micro Levels

The co-knowledge production, as a leverage point in operationalizing resilience, is crucial for **linking macro and micro levels including policy-community linkages.** This is relevant to operational gaps identified in Sect. 2.4.2. From this perspective, operational keys in building/enabling co-knowledge production are listed below (regarding the specific design of co-knowledge production, please see Chapter 4):

Operational Key 1: Components related to the **linkage or process perspectives** need to be incorporated in relevant systems or schemes. **The linkage perspective**-related components include the following: trust, face to face relations, interactive communication, and linkage among different dimensions, actors, organizations or sectors, and between operations/fields and decision-making, and among resources and coordinating functions. **The process perspective**-related components include review and learning processes, self/civic engagement processes, involvement processes for multi-stakeholders, processes for open information, and co-knowledge production processes (see Table 2.3).

Operational Key 2: Operational Key 1 will be further strengthened through **time/temporal perspectives,** such as short, mid, long-term, and intergenerational perspectives, and **scale perspectives**, such as **rescaling scopes and recombining resources** (see Table 2.3). Moreover, in planning or designing a new relevant

Table 2.3 Key components in a resilience approach (Shimizu & Clark, 2019) in considering details and the whole in a continuum

Linkage	Time
• Face to face connections • "Systems" approach • Interactions among sectors and institutions and hub incubator function • Linkage of financial, operational and decision-making • Analytic and integrative approach through multi-dimensions	• Quick response based on daily actions • Multi-temporal (Short, Mid, Long-Term) monitoring-based approach • Intergenerational • Developing new ideas learning from the past • Not depending just on routine/traditional routes and knowing exceptions
Process	Scale
• Evaluation-learning process • Self/Civil society participation • Involvement of multi-stake holders • Inclusion of diversity • Open information • Co-knowledge production system Leverage Point	• Alternative methods • Rescaling • Recreating negative into positive ideas • Recombination of resources for better use (Innovation)

project/program, it is critical to **start from a small scale to accumulate the actions by scaling up with the emphasis of observation, prototyping, and testing or soliciting feedback.**

Operational Key 3: Operational keys 1 and 2 require a **boundary review**. The boundary review refers to paying special attention to **the relationships between issues, systems, and sectors, as well as externalities related to the boundaries.** More specifically, boundary-related areas include **information/data, experience, and lessons learned/insights beyond conventional expertise**, or **organizational or geographical boundaries.** In simpler terms, a boundary review is concerned with **the following interfaces**: person to person, organization to organization, government to communities, government to science, or individuals-community-organization-local-national-global.

Operational Key 4: Based on Operational keys 1–3, it is critical to review all operations **from the macro through the micro levels (macro to micro, or micro to macro) by examining the details and the whole in a continuum and by scaling up and down (or zooming in and out) the relevant scopes** in a non-linear manner through the triple-loop learning process (see Sect. 2.4.2 (6)). Based on these results, it is critical to **update the operations.**

Operational Key 5: Based on Operational keys 1–4, the **"missing links"** will be identified through different dimensions. The missing links can be differentiated from policy gaps as the missing links can be identified by associating policy challenges with lessons learned, experiences and different kinds of knowledge beyond expert knowledge including local knowledge, and indigenous knowledge. In other words, missing links will be identified only through collaborations with different stakeholders, by linking macro–micro actions or policy and communities which experience specific challenges. By addressing the missing links using Operational keys 1–5 through collaboration, the operational gaps composed of "missing links" will be narrowed and directed toward problem-solving oriented actions for transformation.

Operational Key 6: **"Enabling environments" must be ensured for all stakeholders to engage in** building the co-knowledge production/collaborative knowledge production system. This aspect is particularly critical for ensuring special care **for vulnerable populations and communities, given that the systemic challenges surrounding SDGs are often unduly experienced by vulnerable populations and communities.**

The critical points regarding these operational keys can be translated into narrative ways (see Chapter 3) to convey the views or perspectives to different stakeholders as given below:

- Grasping "commons"
- Connecting dots to lines around yourself
- Looking at the details as well as the whole—in other words, *looking at the trees and the forest simultaneously* (a metaphor)
- Looking at both inside and outside
- Paying attention to the hidden dimension or invisibility

- Looking at linkages in the past, present, and future
- Look for boundaries by lowering your walls
- Shifting the axis to look for alternative ways
- Finding the "missing links" and thinking about how to address the "missing links"
- Setting a dialogue-based scheme that enable the all the above aspects.

2.5 Conclusion

Reflecting on Sect. 2.2.3 on "Operational Gaps," (1) actions based on awareness of linkages of SDGs, (2) actions related to macro–micro level linkages, and (3) actions based on paying attention to living systems for transformation, the resilience approach detailed in Sect. 2.4 addresses all of the gaps.

Specifically, regarding (1), the critical four dimensions in enabling resilience, that is, linkage, process, temporal (time), and scale views and their specific components identified above as critical components of the resilience approach, indicate that the different dimensions of linkages related to SDGs are beyond the "issue to issue" linkages of SDGs. Moreover, the resilience approach described here focuses on linking the perspectives to actions, especially by building co-knowledge production systems.

Regarding (2), since a core pillar of the resilience approach is building/enabling co-knowledge production systems, which is a leverage point in operationalizing resilience and is crucial for linking macro and micro levels, the resilience approach can be applied specifically to macro–micro levels, including policy-community linkages.

Regarding (3), that is, "living systems," the resilience approach contributes to developing capacities for "living systems" for mutual interaction and linking those to collective actions; the resilience approach focuses on "looking at natural, human, and social systems in a continuum", or "looking at nature/ecology, different kinds of humans and communities/societies in a continuum," as well as the capacity of natural, human, and social systems to change, adapt, recover, and transform.

Furthermore, the resilience approach covers all dimensions of the pillars of the systemic approach in the context of SDGs (as identified in Chapter 1); this includes the following basic but critical aspects: (a) looking at natural, human, and social systems in a continuum, (b) a system-wide reorganization across different systems/subsystems with a focus on their structures, relationships, and functions by addressing linkages and boundaries in systems/subsystems, and (c) consistent perspectives for different stages for transformation. There are two reasons for this: conceptually, a resilience approach is based on the systems approach, and also practically, the past disasters and lessons learned are largely associated with the systems approach. Thus, the resilience approach is crucial for addressing systemic challenges at local through global levels.

As such, the resilience approach can be applied at different levels, regardless of whether it is macro, meso, or micro levels, especially with regard to linking macro-(meso)-micro levels. The systemic challenges, and especially the operational challenges we face in implementing SDGs, cannot be solved from either the top-down or the bottom-up approach; in other words, we need intersectional and interactive approaches, linking them for collective actions, and the resilience approach can play a major role in this regard.

References

Allen, C., Metternicht, G., & Wiedmann, T. (2018). Initial progress in implementing the Sustainable Development Goals (SDGs): A review of evidence from countries. *Sustainability Science, 13*(5), 1453–1467.

Allen, C., Metternicht, G., Wiedmann, T., & Pedercini, M. (2019). Greater gains for Australia by tackling all SDGs but the last steps will be the most challenging. *Nature Sustainability, 2*(11), 1041–1050.

De Oliveira, A., & Kindornay, S. (2021). *Progressing national SDG implementation: An independent assessment of the voluntary national review*. Reports submitted to the United Nations High-level Political Forum in 2020. Cooperation Canada, Ottawa.

Deubelli, T. M., & Mechler, R. (2021). Perspectives on transformational change in climate risk management and adaptation. *Environmental Research Letters, 16*(5), 053002.

Folke, C. (2016). Resilience (republished). *Ecology and Society, 21*(4), 1–30.

Food and Agriculture Organization of the United Nations. (2020). *A fresh perspective*. Retrieved from https://www.fao.org/forest-resources-assessment/2020/en/. Accessed February 21, 2022.

Fung, J. F., & Helgeson, J. F. (2017). *Defining the resilience dividend: Accounting for co-benefits of resilience planning* (Technical Note 1959). National Institute of Standards and Technology. https://doi.org/10.6028/NIST.TN.1959

Grainger-Brown, J., & Malekpour, S. (2019). Implementing the sustainable development goals: A review of strategic tools and frameworks available to organisations. *Sustainability, 11*(5), 1381.

Helgeson, J., & O'Fallon, C. (2021). Resilience dividends and resilience windfalls: Narratives that tie disaster resilience co-benefits to long-term sustainability. *Sustainability, 13*(8), 4554.

Hillgren, P. A., Seravalli, A., & Emilson, A. (2011). Prototyping and infrastructuring in design for social innovation. *CoDesign, 7*(3–4), 169–183.

Hoff, H. (2011). *Understanding the nexus*. Background Paper for the Bonn 2011 Conference: The Water, Energy and Food Security Nexus. Stockholm Environment Institute, Stockholm.

Griggs, D. J., Nilsson, M., Stevance, A., & McCollum, D. (2017). *A guide to SDG interactions: From science to implementation*. International Council for Science.

Holling, C. S. (1973). Resilience and stability of ecological systems. *Annual Review of Ecology and Systematics, 4*, 1–23.

IUCN. (2016). *A resolution at the 2016 World Conservation Congress* (WCC-2016-Res-069). Retrieved from https://portals.iucn.org/library/sites/library/files/resrecfiles/WCC_2016_RES_069_EN.pdf. Accessed February 21, 2022.

Jackson, S. (2010). *Architecting resilient systems: Accident avoidance and survival and recovery from disruptions*. Wiley.

Lim, M. M., Jørgensen, P. S., & Wyborn, C. A. (2018). Reframing the sustainable development goals to achieve sustainable development in the anthropocene—A systems approach. *Ecology and Society, 23*(3).

Mayrhofer, J. P., & Gupta, J. (2016). The science and politics of co-benefits in climate policy. *Environmental Science & Policy, 57*, 22–30.

New York Declaration on Forests (NYDF)Assessment Partners. (2021). *Taking stock of national climate action for forests: 7th progress report*. Retrieved from https://forestdeclaration.org/wp-content/uploads/2021/10/2021NYDFReport.pdf. Accessed on January 2022.

Okada, N. (2021, November 26). *Another challenge: Systemic thinking and design for community-based/humans-focused disaster risk governance.* Invited speech a DPRI SOGO-Bosai seminar 50th Session.

Pahl-Wostl, C. (2009). A conceptual framework for analysing adaptive capacity and multi-level learning processes in resource governance regimes. *Global Environmental Change, 19*(3), 354–365.

PBES. (2019). *Summary for policymakers of the global assessment report on biodiversity and ecosystem services of the Intergovernmental Science-Policy Platform on Biodiversity and Ecosystem Services* (S. Díaz, J. Settele, E. S. Brondízio, H. T. Ngo, M. Guèze, J. Agard, A. Arneth, P. Balvanera, K. A. Brauman, S. H. M. Butchart, K. M. A. Chan, L. A. Garibaldi, K. Ichii, J. Liu, S. M. Subramanian, G. F. Midgley, P. Miloslavich, Z. Molnár, D. Obura, A. Pfaff, S. Polasky, A. Purvis, J. Razzaque, B. Reyers, R. Roy Chowdhury, Y. J. Shin, I. J. Visseren-Hamakers, K. J. Willis, & C. N. Zayas, Eds.). IPBES Secretariat, Bonn, Germany. Retrieved from https://ipbes.net/global-assessment. Accessed on February 21, 2022.

Plattner, H. (2010). *An introduction to design thinking.* Process Guide. https://dschool-old.sta nford.edu/sandbox/groups/designresources/wiki/36873/attachments/74b3d/ModeGuideBOO TCAMP2010L.pdf. Accessed March 10, 2022.

Senge, P. M., Scharmer, C. O., Jaworski, J., & Flowers, B. S. (2005). *Presence: An exploration of profound change in people, organizations, and society.* Currency.

Shimizu, M. (2015). *Collaborative knowledge creation based resilience.* Kyoto University Press. (in Japanese).

Shimizu, M., & Clark, A. (2019). *Nexus of resilience and public policy in a modern risk society.* Springer.

Tanner, T. M., Surminski, S., Wilkinson, E., Reid, R., Rentschler, J. E., & Rajput, S. (2015). *The triple dividend of resilience: Realising development goals through the multiple benefits of disaster risk management.* Global Facility for Disaster Reduction and Recovery (GFDRR) at the World Bank and Overseas Development Institute (ODI), London. www.odi.org/tripledividend

UN Climate Change Conference UK. (2021). *Glasgow leaders' declaration on forests and land use.* Retrieved from https://ukcop26.org/glasgow-leaders-declaration-on-forests-and-land-use/. Accessed February 11, 2022.

United Nations Office for Disaster Risk Reduction (UNDRR). (2021). *Words into action: Nature-based solutions for disaster risk reduction.* Retrieved from https://www.undrr.org/words-action-nature-based-solutions-disaster-risk-reduction. Accessed February 11, 2022.

Weitz, N., Carlsen, H., Nilsson, M., & Skånberg, K. (2018). Towards systemic and contextual priority setting for implementing the 2030 agenda. *Sustainability Science, 13*(2), 531–548.

Wild, T., Freitas, T., & Vandewoestijne, S. (Eds.). (2020). *Nature-based solutions: State of the art in EU-funded projects.* Publications Office of the European Union. Retrieved from https://oppla.eu/product/21302. Accessed on February 11, 2022.

Zhongming, Z., Linong, L., Xiaona, Y., Wangqiang, Z., & Wei, L. (2021). *Nature-based solutions in Europe: Policy, knowledge and practice for climate change adaptation and disaster risk reduction.* Publications Office of the European Union. Retrieved from https://www.eea.europa.eu/publicati ons/nature-based-solutions-in-europe. Accessed February 11, 2022.

Mika Shimizu is an Associate Professor in Graduate School of Advanced Integrated Studies in Human Survivability, Kyoto University. Her long years' experiences as a policy researcher in East-West Center in Washington DC and Honolulu, Hawaii in the United States greatly contributed to publishing this book. She holds an M.A. from American University and a Ph.D. in International Public Policy from Osaka University (2006). She has been extensively involved in interdisciplinary and transdisciplinary research projects related to disasters/infectious diseases, sustainability, and climate change issues with the focus on resilience. Her major publications include *Nexus of Resilience and Public Policy in a Modern Risk Society* (Co-Author: Allen Clark, Springer, 2019).

Part II
Underlying Themes: Global to Local Linkages

Chapter 3
Reconciling Risk, Resilience, and Sustainability: Learning from Narratives

Ilan Chabay ⓘ

Abstract The continuing unsustainable pattern of human behavior creates enormous risks of increasingly frequent and severe disasters impacting the complex social-ecological-economic-technical system (SES) in which all life on Earth is embedded. This chapter outlines key concepts of complex systems as the foundation for understanding resilience, sustainability, and acute or chronic disruptive stressors of the SES. The relationship between resilience and sustainability has the potential for synergies or tradeoffs, depending on normative judgements particular to the cultures and contexts in which the risks may manifest. Important insights into the normative landscape in co-located or online communities can be gained by collecting and analyzing their narratives of imagined futures and social identities, which is one of the key research areas of the international Knowledge, Learning, and Societal Change Alliance (KLASICA).

3.1 Introduction: Current Crises and Systemic Risks

Long after the first reported outbreak, we are still living in the throes of a devastating pandemic as genetic variants of the SARS-Cov-2 virus cause waves of COVID-19 infections sweeping across every country and corner of our planet. In addition to the tragedy of so many lives lost, there is also the tremendous disruptions to the health and well-being of society with negative impacts on health service sectors, economic viability of businesses, supply chains, education, democratic institutions, and equity for different populations. Nearly every part of the global system in which we live has been affected.

These multiple impacts are not simple coincidences. They are consequences of the interactions in the complex socio-ecological-economic-technical system (SES) in which all life on Earth is embedded. We humans are an integral part of the global to local complex SES on Earth. As is the case with all living species, we sustain

I. Chabay (✉)
Institute for Advanced Sustainability Studies (IASS), Potsdam 14467, Germany
e-mail: Ilan.Chabay@iass-potsdam.de

ourselves by exploiting the ecological systems on which we are totally dependent. However, it is only the human species that has increasingly extensively and profoundly changed Earth's environment. This enormous human impact on Earth is captured in the concept of the Anthropocene Era (Crutzen, 2002). The critical and indeed existential challenges we have created and are facing in the Anthropocene Era span local to global scales. Crucially for implementation, they are distributed in various manifestations across the world and range from the present to the times of future generations.

Meeting these challenges requires both resilience and sustainability of the SES in the present and in the future. In turn, this requires our understanding of the multiple temporal and spatial scales of systems, boundaries, and the interactions among the interdependent components. How can we become better able to avoid or diminish the risk of devastating impacts and manage effective responses to hazardous events that threaten our societies? At the same time, how can we develop alternatives to the paths and path dependencies that led to such profound systemic risks?

The answers to these questions require us to find resilient and sustainable solutions which are implementable in each of the local bio-geo-physical and cultural contexts, such that the totality of these local and regional solutions forms a globally coherent approach, as is embodied in the United Nations Sustainable Development Goals (SDGs) (United Nations, 2015). In addition to the attention to science and policy at the global level, to achieve sustainability, it is essential to devote more attention to socio-cultural dynamics, modeling, and implementation at many smaller spatial–temporal scales (Chabay et al., 2021).

This chapter begins with an overview of systemic risks to lay the ground for a discussion of how resilience and sustainability are interrelated and how they differ as responses to systemic risks, specifically those systemic risks arising from unsustainable patterns of human behaviors. In the latter part of the chapter, I argue that narratives provide unique, valuable insights into perceptions, visions, and social identities that help in reconciling tradeoffs and identifying synergies between resilience and sustainability of SES.

3.2   Complex Systems and Systemic Risks

Resilience and sustainability are both generally desirable for any community, especially when it faces the potential of shocks and stressors. They are interdependent dynamic properties of the complex SES in which they function. To understand what this entails and how the two interrelated properties can be mutually reinforcing or negatively interfering, an overview of the concepts of complex systems and systemic risk is useful to establish a common understanding of the concepts. With these concepts in mind, we can consider the interplay of resilience and sustainability in the complex socio-economic-ecological-technical system of a community.

What constitutes a system is defined by what is contained within its boundaries (e.g., animal cell membranes which enclose the cytoplasm and are selectively permeable to certain molecules and ions). Systems can be deterministic or stochastic and complicated, or they may be complex.

3.2.1 Complex Systems

1. contain multiple interacting components that
2. exhibit feedback loops connecting interdependent components in space and time.
3. Feedback may be non-linear and either positive (amplifying) or negative (damping),
4. emergent properties arise from multiple interdependent components, leading to

   a. inherent uncertainty, which cannot be eliminated,
   b. ambiguity due to human value judgements and contingency of human decisions,
   c. high probability of unintended or unanticipated consequences of actions because of interdependencies among components and because the relationship between events and their cause(s) may not be apparent,
   d. the possibility of exceeding tipping points, causing irreversible system change

Systemic risks are characterized by five attributes (Schweizer et al., 2021): a high degree of complexity in their causal or functional relationships; transboundary effects; stochastic distribution of effects; high degree of uncertainty; and nonlinearities that can lead to irreversible change at tipping points. They are also associated with cascades of impacts within the affected domain or beyond (known as "ripple effects") (Renn et al., 2020). These ripple effects are linked to and exacerbated by processes of social amplification or attenuation of risk shaped by social perceptions and feedback beyond the subsystem in which the risk originated (Renn et al., 1992). Spatial and temporal dispersion of consequential effects, as well as the multitude of intervening variables between causes and effects make it very difficult to identify the causes of triggers, consequences, feedback loops, and impacts of systemic risks (Lucas et al., 2018; Renn et al., 2019).

Multiple intervening factors that mutually affect one another and attenuate or amplify causal relationships are characteristic of complex systems. This level of complexity creates uncertainty beyond the expected normal statistical distribution of effects. Identical causes may lead to different effects, depending sensitively on the initial situation of a systemic risk. This feature is known from multi-agent, multi-moderator risks, such as the effects of chemical mixes on human health and from studies of chaotic phenomena (Strogatz, 2014). Furthermore, with systemic risks, there is high uncertainty about both magnitude and probability of expected adverse effects and propagation of impacts through the system. Probabilities and distributions of occurrences may change substantially due to non-linearity with extreme sensitivity to initial conditions making it difficult to extrapolate from past distributions to predict future events.

Systemic risks may be endogenous or exogeneous with respect to the affected domains. They may originate in the environment, human-made systems, or biological systems, but they may have causes and impacts that span boundaries or sectors where their ripple effects have a large or small impact (Aven & Renn, 2020). The different impacts of the cascading effects of systemic risks are conditioned by different local contexts in which they occur. Systemic risks can transcend boundaries of jurisdiction, nationality, or sectoral responsibility and therefore often call for multilevel governance (Hooghe & Marks, 2001) and international cooperation. The COVID-19 pandemic demonstrates these transboundary effects of systemic risks rather well. It also dramatically illustrates the deep interdependence of subsystems of the complex global SES. This can be seen in the interactions between humans and animal reservoirs of viruses, rapid viral distribution through global trade expansion, impact of widespread illness on extended supply chains, and simultaneous near or total collapse of health systems and hospitals in multiple locations.

In addition, many systemic risks have the potential to lead to phenomenological or societal tipping points (e.g., decreases in habitat extent and connectivity leading to species extinction or drought impacting food availability with consequent price increases adding to social tension and leading to an uprising and political regime change, respectively). The system changes irreversibly once a tipping point has been reached. These changes in condition may even include a complete collapse of the system (Lenton et al., 2008). The developments leading up to a tipping point may be "quietly" incremental and often go undetected. However, once a tipping point has been reached, it is too late to adapt and reverse the consequences.

Even when a clearly detectable earthquake and tsunami or a massive rainfall from hurricane or typhoon occurs, a warning system may provide only a very short time prior to the triggering event for inhabitants in the affected area to react and activate protective measures. However, some early warning systems in operation together with improved education and training have been effective in increasing the resilience of regions (Kéfi et al., 2014).

In the case of growing societal stress before a tipping point is reached, the problem is detecting the increase in these chronic stress conditions and anticipating how the impact on people in the affected area might be amplified or diminished by the perception of the event (Renn et al., 1992). Detecting the accumulation of stress is difficult because people often become habituated to the stress or because the cause of the stress is masked by other patterns (e.g., stigma associated with a lifestyle or marginalized group that encourages the society at large to ignore the buildup of tension). Such slow changes in conditions may well remain "below the radar" until the system changes irreversibly into a different condition that adversely affects the well-being of some or all inhabitants in the area. Whether threatened by exogenous sudden shock or by approaching a tipping point, building resilience, and finding appropriate sustainable solutions are essential to maintaining the core functions and fundamental well-being of the threatened community and often that of many other interdependent communities and regions.

3.2.2 Resilience: A Dynamic Property of the System

Resilience can be defined as the capacity of the affected SES to avoid, absorb, or adapt to a major shock to the system without losing its essential functions, processes, values, and relationships that lend meaning to human society and support a healthy ecosystem. This is a dynamic response, such that if a threat cannot be avoided, the system acts to adaptively preserve highly valued elements and functions of the society. Resilience includes the ability to reduce initial adverse effects (*absorptive capability*) of a disruptive shock and the time and costs needed for the system to return to functionality (*adaptive and restorative capability*). Note that disruption may occur as *acute* shocks or chronic stress and can have endogenous or exogenous origins (Aven et al., 2018). Resilience of the complex SES is represented in simplified form as one slice through a multi-dimensional surface in Fig. 3.1.

In Fig. 3.1, a system (red) is displaced from its equilibrium state S_2 by a major shock (e.g., earthquake and infrastructure collapse) that sends it toward transition state TS_{23}. Depending on the power and impacts of the shock, the system may be driven past the transition state TS_{23} and into a new configuration (blue) with significantly different characteristics than it had in S_2. Alternatively, with sufficient resilience, the system may return in a short time to a close resemblance of the original equilibrium state S_2 in which case, the critical functions of the system are maintained or restored. The resilience of the system in S_2 is roughly represented by the difference

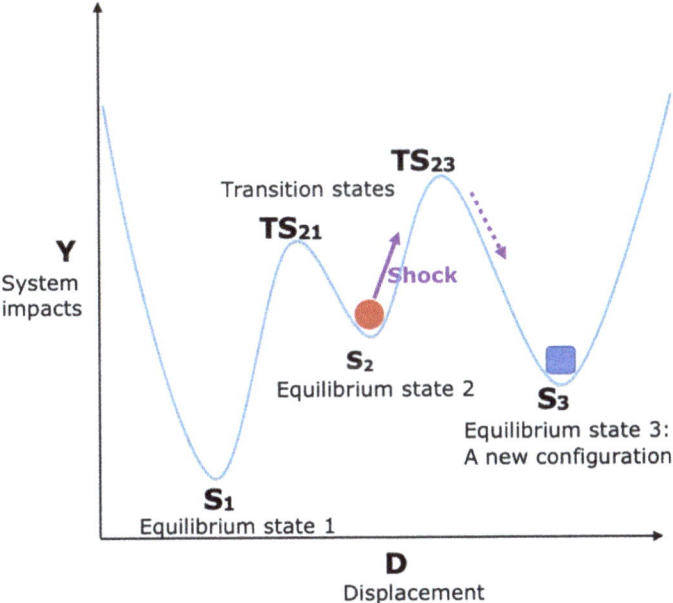

Fig. 3.1 System Impacts, Equilibrium States, and Resilience

in Y between S_2 and a transition state, i.e., the "depth" of the minimum. The displacement D represents the change in a system property, such as number of houses and businesses flooded and Y represents the impacts of the displacement, e.g., number of people left homeless or without being able to earn livelihoods or at risk of water borne disease. A different form of shock and initial conditions (e.g., drought, famine, and armed revolt) might have driven the system from S_2 to S_1.

Whenever a shock to a SES occurs, the community's resilience is tested. Building resilience prior to a potential shock (e.g., by investing in fire-fighting equipment and training firefighters in areas of high risk of wildfires near homes) increases the height of the transition states and simultaneously invests in benefits for the community, which can yield resilience dividends (Ahmed, 2017; Fung & Helgeson, 2017). But preparation may not be sufficient to avoid a disaster completely if the shock is powerful, affects multiple interdependent components of the SES, spans multiple governance jurisdictions, or when planning and training for the type of shock is incomplete or inadequate. Planning and training needed for building resilience to shocks depends on knowledge of the likelihood of occurrence in a specified timeframe and the potential spatial and temporal scale and power of the shock, and its likely impacts on components of the SES. It also depends on available resources to understand and track these issues, as well as available resources (money, time) to implement resilient capacity and planning prior to an event. Predicting future conditions of a complex system to support improved resilience can only be obtained through foresight techniques, including scenario building and modeling. However, predictions are inevitably accompanied by substantial uncertainty and ambiguity of values and interpretation. Computer scientist Alan Kay remarked in 1971 that "the best way to predict the future is to invent it." If one includes social as well as technological invention in the spirit of that remark, the concept of sustainability inspires the effort to invent and implement pathways to a better future for present and future societies under changing conditions and within the limits of critical resources on Earth.

3.2.3 Sustainability: Pathways to just and Equitable Sustainable Futures

It has become increasingly urgent and critical to design and implement new pathways out of our present unsustainable patterns of living that put us at risk of transgressing the limits of critical natural resources, as described in the Planetary Boundaries papers (Rockström et al., 2009; Steffen et al., 2015) and equally of transgressing societal boundaries and perpetuating or exacerbating societal inequities (Leach et al., 2013). This challenge is addressed by the effort to fulfill the UN Sustainable Development Goals (SDGs) (United Nations, 2015). The SDGs and the associated targets represent a remarkable step forward by the nations of the world in establishing aspirational agreement and calling for national targets to achieve the goals. Though crafted as global goals with individual national implementation targets, it is at local and regional scales that sustainability targets can inspire visions of sustainable futures and lead to commitment to actions in the different contexts and cultures across the world.

Central to the concept of sustainability is the recognition that humanity is an integral part the local, regional, and global SES in which we humans live and on which we are entirely dependent. As summarized by Olsson et al., "(Folke et al., 2011) argue that any attempts to create sustainability transformations should involve strategies for 'reconnecting to the biosphere,' which entails a view of humans and nature as an integral whole within which a healthy planet is the premise for economic and social development." (Olsson et al., 2014, p. 2).

Sustainability has been defined in various ways since coming to public prominence in the 1987 Brundtland Commission report for the UN that called for "development that meets the needs of the present without compromising the ability of future generations to meet their own needs" (World Commission on Environment and Development (UN), 1987). The term "development" is often contested, which is why I prefer to use "sustainable futures" to emphasize both the plurality of multi-generational and multi-species futures in different cultures and contexts and the idea that sustainability is predicated on the viability over extended time of interdependent social, ecological, economic, technical subsystems of the overall planetary SES. Past and current events clearly indicate that the viability of SES at multiple spatial and temporal scales across the world is subject to increasingly severe chronic and acute stress in large part due to unsustainable societal behavior patterns. Among the many stressors are pandemics, climate change, wildfires, drought, floods, landslides, aging populations, loss of local expertise in agriculture, massive migration of at-risk populations, loss of biodiversity, and loss of traditional livelihoods, cultures, and languages. These stressors can ultimately lead to tipping points and catastrophic events, which in turn places our reliance on resilience to avoid or mitigate the impact of the events as discussed in the preceding section. Sustainability and resilience are intertwined through the societal choices, policies, and actions that influence future conditions to which the Earth's SES will be exposed (Shimizu & Clark, 2019).

Redman (2014) contrasts sustainability and resilience in the following passage: "The strength of a sustainability approach is that it systematically examines future options, assigns values to those options via indicators, and customizes its strategies to attain those options. It rigorously integrates normative values and anticipatory thinking into a scientific framework (Clark & Dickson, 2003; Swart et al., 2004). In contrast, the strength of a resilience approach is that it develops adaptive capacity and/or robustness into the system so that the system can gracefully weather the inevitable, but unspecified, system shocks and stressors. Resilience approach does not require predicting outcomes. Instead, it builds social and natural capital and enhances adaptive capacity to cope with unknown futures (Carpenter & Folke, 2006; Folke et al., 2010). Simply put, sustainability prioritizes outcomes; resilience prioritizes process" (Redman, 2014). Unsaid in this view is that sustainability is an emergent property, not a singular outcome, and resilience is also a property of the system built through processes.

The relationship between resilience and sustainability carries with it two conundrums. One is potential tradeoffs. For example, tradeoffs may arise due to the differences in intergenerational priorities. As discussed above, resilience is the capacity to avoid, mitigate, or cope with events when they occur and, in the period immediately following events, recover and build back better than before (Okada, 2021) based upon the assessment of possible future hazardous conditions. Sustainability deals with identifying and implementing socio-ecologically constructive pathways from current to future conditions and carries the potential to decrease the likelihood or severity of catastrophic events occurring. Thus, accomplishing steps toward long term goals (e.g., SDGs) can diminish the likelihood of hazards becoming actualized and decrease the impacts if a shock occurs. This in turn decreases dependence on resilience needed to avoid or cope with the shock and its impacts.

However, a community may also face tradeoffs between expending resources for improving resilience regarding a particular type of hazard in the short term or expending resources to alter longer term trajectories of the SES, such that the hazards are eliminated or significantly diminished. "Synergies are the interactions between resilience and sustainability, such that their combined effects are greater than the sum of their effects if implemented separately. Tradeoffs are a balancing of resilience and sustainability when it is not possible to execute on both fully and at the same time (due to financial or other constraints)." (Helgeson & O'fallon, 2021, p. 3).

Saunders and Becker address the tradeoff in the context of the 2010 and 2011 series of earthquakes in Canterbury, New Zealand. They write "…if events keep occurring and impacts start compounding, then short term adaptive measures (e.g., insurance, rebuilding) may not address the hazard problem effectively. Instead, a community may need to employ a set of adaptive measures that are more useful in longer term (e.g., retiring land, zoning)—something that is more in line with the concept of sustainable development." (Saunders & Becker, 2015).

In many cases, difficult decisions on tradeoffs must be made. An example of a survey method that "quantifies and compares the relative resilience across watershed systems and potential trade-offs among different aspects of the social-ecological system, e.g., between social, economic, and ecological contributions" is discussed by Allen et al. (Allen et al., 2018). The tradeoffs in such watershed systems are both about resilience in the watershed area and sustainability of the resources fed by the watershed. Any choices therefore impact the social, ecological, and economic components of the sustainability differently. In a different study of watersheds, the authors write that "the social and economic aspects of the Platte Basin are highly desirable, and, unconsciously at least, a decision has been made to sacrifice the ecological component in favor of the social and economic components." (Birge et al., 2014).

The second conundrum is the difference regarding normativity. Elmqvist et al. delineate the difference in the following: "Resilience is fundamentally non-normative and an attribute of the system and applicable to different subsystems", whereas "sustainability is a normative concept, representing the vision for society." (Elmqvist et al., 2019, p. 269). This statement may be taken as broadly useful in contrasting resilience and sustainability. However, it can be challenged in that resilience describes the capacity to return to a condition in which the essential functions of the system are restored, which leaves open the normative judgement of what are essential functions and which values prevail in the social system.

Normative judgement and aspiration are at the very core of sustainability as an emergent property of SES. How are those normative aspects of sustainability perceived, understood, and communicated in societies? Narratives are an important means of communicating normative visions and motivating actions for resilience and sustainability. Though narratives have been studied in great depth in literary and historical contexts, their value as sources of significant insights into social identities, future visions, and societal dynamics in scientific studies, including of resilience and sustainability, has only recently come to the fore (Bremer & Funtowicz, 2015; Chabay, 2015; van der Leeuw, 2019; Veland et al., 2018).

3.2.4 Narratives: Normative Expressions of Visions and Identities

Throughout human existence, narratives have been a fundamental form of human communication with which to express social identities, reinforce culture, reflect or influence individual and group behaviors, and frame imagined futures (Bruner, 1991; Morgan & Wise, 2017). Narratives generally contain stories with actors, scenes, plot, but crucially they are also purposeful and fulfill three functions (Chabay et al., 2019):

1. they structure, prioritize, and ascribe meaning to experiences and beliefs (e.g., experiences and ways of living of the Inuit learned over centuries allow them to survive in the harsh conditions of the Arctic)
2. provide orientation for facing uncertain and unfamiliar contexts (e.g., vaccines have been used very successfully for many other communicable diseases, so we know why vaccinations for Covid-19 are critical for protecting society)
3. facilitate sense making and decision making in highly complex social–ecological systems by representing core values (norms) and ideas with a simplified, condensed, and emotionally engaging way in which only a few specific properties of characters and their environment are emphasized, thus becoming more memorable or iconic (e.g., the protagonist in the movie eventually recognizes the source of her fear that has prevented her from fulfilling her needs and thus becomes able to overcome the forces arrayed against her (Maggs & Chabay, 2022)

Narratives influence the cohesion and resilience of a community, especially under perceived threat and actual events from exogenous or endogenous hazards or accumulating stress, because they provide a concisely communicable, recognizable, and memorable expression of a vision of a desirable future (or fear of a dystopian future) and a sense of shared identity. They represent "imagined futures" (Beckert, 2013) or alternative ways of conceptualizing what has yet to happen (Chabay et al., 2019).

Such narratives may shift social norms and attitudes and form a basis for decisions by individuals and groups.

By highlighting one's identity as an individual in the community *vis-à-vis* the identities of the relevant community, these narratives provide strong motivation for decisions and actions that are perceived to be in the interests of the individual and self-selected community. These narratives are of the sort that can be considered "cultural and identity-protective cognition" expressions (Kahan, 2017; Kahan et al., 2007).

They reflect and may influence the sense of agency and responsibility individuals feel about acting on an issue. The connection to group identity provides a sense of collective agency beyond any individual's sense of personal capacities. This can play a very important part in strengthening resilience of the community, especially when members of the community may have to act in the immediate circumstances of a disaster without benefit of external information or direction, as noted in the three functions above.

The efforts to accomplish the SDGs by the target date of 2030 are also subject to influences of narratives in circulation in different communities and transmitted between communities. Building momentum toward achievement of the SDGs across the world and moving on pathways toward sustainable futures at the community and regional scales requires not only individual, but collective behavior change in different contexts and cultures.

Narratives influence collective behavior changes in different situations, such as protecting natural resource commons (fish stocks) in Malawi and Micronesia and rebuilding indigenous food culture to decrease reliance on food imported from France in Martinique, French West Indies (Chabay et al., 2019). In these different cultures, the common elements that led to collective behavior change to more sustainable outcomes were on one hand a narrative of a desirable change for the future of the community and on the other hand, a narrative that reinforced the sense of belonging to the community and having a shared social identity.

Narrative expressions of underlying worldviews of groups or communities, provide insight into social cohesiveness or polarization, perceptions and attitudes to policies or regulations, and indicate group beliefs, values, and rationales for their views (Glynn et al., 2017). The links between narratives, politics, and pathways to sustainability were presciently discussed for the case of governance of epidemics by Leach, et al. (Leach et al., 2010). These links are now increasingly evident as the COVID-19 pandemic is sorely testing our socio-environmental systems' ability to deal with massive and prolonged stresses. The critical challenges we face are due to systemic risks of unsustainability that have led and will lead to crises that cannot be solved in a linear, piecemeal fashion (Renn et al., 2020), but will require holistic, systemic solutions (Lucas et al., 2018).

Narratives circulate in multiple forms in communities, including through conversation, dance, music, speeches, news media, and digital social media. Especially in the rapidly expanding pervasive - though still neither ubiquitous nor equitably distributed - domain of many-to-many communication in digital media, many narratives are constantly exchanged by members of online communities (Helgeson et al.,

2022). These narratives often evolve rapidly in asynchronous exchange and may frag-
ment into separate communication channels used by in-group members who share a
sense of common social group identity or common purpose, thus forming exclusive
polarized echo chambers and affecting social norms, politics, and policies.

An example of the power of narratives to influence and justify policy and the
subsequent negative impact on the resilience of society can be seen in the story
of two ice storms in Texas, USA that occurred in 2011 and 2021. In both storms,
millions of people were left without electrical power for several days. "...the first
major blackout in Texas in 2011 did not result in any meaningful changes in infrastructure to create resiliency to a similar event in the future, nor have changes been
evident in the aftermath of the 2021 blackouts." (Towers et al., 2022). A libertarian
narrative reinforced in a large political and corporate echo chamber in Texas decried
dependence on US federal regulation and led to the state policy that left the Texas
power grid decoupled from the US national power grid. Consequently, a system for
backup from the national grid was not available during the storms to protect from
the blackout. At the same time, a false narrative to protect the power companies was
promoted by the governor of Texas and others in politics and corporate sector that
blamed the blackout on icing of windmills and solar panels, though the renewables
supplied only a small fraction of the power demand of Texas and therefore their
failure did not contribute substantially to the blackout.

Collecting and characterizing narratives in circulation among different communities is an important means of detecting and analyzing the prevailing range of perceptions, visions, and fears or hopes in communities of interest (Chabay et al., 2019;
Bakamo Social, 2017). To support and advance this process, a Digital Observatory
of Narratives of Sustainability (DONS) has been recently proposed by (Helgeson
et al., 2022) as an open international platform for collecting and analyzing narratives in digital media. Its purpose is to enable greater understanding of perceptions,
norms, and vision in the political, ideological, cultural, geographical, institutional,
identity, generational and power dimensions of SES in online communities or groups.
Analyzing and addressing the concerns that shape narratives can in turn guide participatory analytic-deliberative dialogues (Renn, 1999) and lead to new civic actions,
laws, regulations, and investments in research and development of new (social and
material) innovations that offer inclusive, just, and equitable pathways for the future.

Understanding and catalyzing collective behavior change on pathways to sustainable futures is the purpose of the KLASICA international research alliance, which
was initiated by Ilan Chabay in 2009 as the Knowledge, Learning, and Societal Change (KLSC) project. It is being expanded and relaunched in 2022 as the
KLASICA. It is an international collaborative research hub and resource platform
for catalyzing innovative thinking, research, and practice through active collaboration among scholars and practitioners in the arts, humanities, social sciences, natural
sciences, medicine, and engineering. KLASICA will contribute to and draw from
narratives in the DONS described above (Helgeson et al., 2022), along with concepts,
methods, and analysis for understanding and facilitating collective behavior change.
It will organize and host thematic KLASICA symposia, coordinate mentoring of
early career scholars and practitioners, and organize and facilitate writing of research

papers, perspective articles, and topical reports to support policies and actions on issues of sustainability.

3.3 Conclusion: Toward Greater Resilience and More Sustainable Futures

In this chapter, I have started with an outline of systemic risk in the context of social-ecological-economic-technical systems (SES) as a basis for understanding the synergy and tension between resilience and sustainability at multiple spatial and temporal scales. I introduced the use of narratives of future vision and social identity to understand factors that hinder or enhance community resilience and movement on paths to sustainability and the SDGs.

Improving and deepening the understanding throughout society of the key characteristics of the complex SES of which we are an integral part and employing valuable insights from narratives of vision and social identity in communities of the world are essential steps in making meaningful, informed choices that will improve resilience and lead toward just and equitable sustainable futures.

Acknowledgements I am grateful for helpful conversations on resilience and sustainability with Associate Professor Mika Shimizu (GSAIS, Kyoto University) and Dr. Pia-Johanna Schweizer (IASS Potsdam, Germany) and for valuable comments on the manuscript from Dr. Jennifer Helgeson (NIST, USA). I also want to acknowledge the kind invitation from Professor Ana Maria Cruz and the support of the Disaster Prevention Research Institute of Kyoto University for a five month long visiting professorship in 2021, which allowed me to engage in and learn from many productive conversations on risk, resilience, and sustainability with faculty and students.

References

Ahmed, S. (2017). The resilience dividend: Being strong in a world where things go wrong. *Resilience, 5*(3), 222–224. https://doi.org/10.1080/21693293.2016.1153775

Allen, C. R., Birge, H. E., Angeler, D. G., Arnold, C. A. (Tony), Chaffin, B. C., DeCaro, D. A., Garmestani, A. S., & Gunderson, L. (2018). Quantifying uncertainty and trade-offs in resilience assessments. *Ecology and Society, 23*(1), art3. https://doi.org/10.5751/ES-09920-230103

Aven, T., Ben-Haim, Y., Andersen, H. B., Cox, T., Droguett, E. L., Greenberg, M., Guikema, S., Kröger, W., Renn, O., Thompson, K. M., & Zio, E. (2018). *Society for risk analysis glossary* (p. 9).

Aven, T., & Renn, O. (2020). Some foundational issues related to risk governance and different types of risks. *Journal of Risk Research, 23*(9), 1121–1134. https://doi.org/10.1080/13669877.2019.1569099

Bakamo Social. (2017). *Summary of the French election social media landscape report 2017 Key findings* (pp. 1–13).

Beckert, J. (2013). Imagined futures: Fictional expectations in the economy. *Theory and Society, 42*(3), 219–240. https://doi.org/10.1007/s11186-013-9191-2

Birge, H. E., Allen, C. R., Craig, R. K., Garmestani, A. S., Hamm, J. A., Babbitt, C., Nemec, C., & Schlager, E. (2014). Social-ecological resilience and law in the Platte River Basin. *Idaho Law Review, 51*(1), 229–256.

Bremer, S., & Funtowicz, S. (2015). Negotiating a place for sustainability science: Narratives from the Waikaraka Estuary in New Zealand. *Environmental Science and Policy, 53*, 47–59. https://doi.org/10.1016/j.envsci.2014.11.006

Bruner, J. (1991). The narrative construction of reality. *Critical Inquiry, 18*(1), 1–21.

Carpenter, S. R., & Folke, C. (2006). Ecology for transformation. *Trends in Ecology & Evolution, 21*(6), 309–315. https://doi.org/10.1016/j.tree.2006.02.007

Chabay, I. (2015). Narratives for a sustainable future: Vision and motivation for collective action. In B. Werlen (Ed.), *Global Sustainability* (Issue December 2012, pp. 51–61). Springer International Publishing. https://doi.org/10.1007/978-3-319-16477-9_3

Chabay, I., Koch, L., Martinez, G., & Scholz, G. (2019). Influence of narratives of vision and identity on collective behavior change. *Sustainability, 11*(20), 5680. https://doi.org/10.3390/su11205680

Chabay, I., Renn, O., van der Leeuw, S., & Droy, S. (2021). Transforming scholarship to co-create sustainable futures. *Global Sustainability, 4*, e19. https://doi.org/10.1017/sus.2021.18

Clark, W. C., & Dickson, N. M. (2003). Sustainability science: The emerging research program. *Proceedings of the National Academy of Sciences of the United States of America, 100*(14), 8059–8061. https://doi.org/10.1073/pnas.1231333100

Crutzen, P. J. (2002). Geology of mankind. *Nature, 415*(6867), 23–23. https://doi.org/10.1038/415023a

Elmqvist, T., Andersson, E., Frantzeskaki, N., McPhearson, T., Olsson, P., Gaffney, O., Takeuchi, K., & Folke, C. (2019). Sustainability and resilience for transformation in the urban century. *Nature Sustainability, 2*(4), 267–273. https://doi.org/10.1038/s41893-019-0250-1

Folke, C., Carpenter, S. R., Walker, B., Scheffer, M., Chapin, T., & Rockström, J. (2010). Resilience thinking: Integrating resilience, adaptability and transformability. *Ecology and Society, 15*(4), 20. https://doi.org/10.5751/ES-03610-150420

Folke, C., Jansson, Å., Rockström, J., Olsson, P., Carpenter, S. R., Stuart Chapin, F., Crépin, A. S., Daily, G., Danell, K., Ebbesson, J., Elmqvist, T., Galaz, V., Moberg, F., Nilsson, M., Österblom, H., Ostrom, E., Persson, Å., Peterson, G., Polasky, S., … Westley, F. (2011). Reconnecting to the biosphere. *Ambio, 40*(7), 719–738. https://doi.org/10.1007/s13280-011-0184-y

Fung, J. F., & Helgeson, J. F. (2017). Defining the resilience dividend: Accounting for co-benefits of resilience planning. https://www.nist.gov/publications/defining-resilience-dividend-accounting-co-benefits-resilience-planning

Glynn, P. D., Voinov, A. A., Shapiro, C. D., & White, P. A. (2017). Earth 's future from data to decisions: Processing information, biases, and beliefs for improved management of natural resources and environments Earth 's future. *Earth's Future, 6*(5), 757–761. https://doi.org/10.1002/eft2.199

Helgeson, J., Glynn, P., & Chabay, I. (2022). Narratives of sustainability in digital media: An observatory for digital narratives *Futures, 142*, 103016. https://doi.org/10.1016/j.futures.2022.103016

Helgeson, J., & O'fallon. (2021). Resilience dividends and resilience windfalls: Narratives that tie disaster resilience co-benefits to long-term sustainability. *Sustainability (switzerland), 13*(8), 4554. https://doi.org/10.3390/su13084554

Hooghe, L., & Marks, G. (2001). *Multi-level governance and European integration*. Rowman & Littlefield Publishers.

Kahan, D. M. (2017). Misconceptions, misinformation, and the logic of identity-protective cognition. *SSRN Electronic Journal*. https://doi.org/10.2139/ssrn.2973067

Kahan, D. M., Braman, D., Gastil, J., Slovic, P., & Mertz, C. K. (2007). Culture and identity-protective cognition: Explaining the white-male effect in risk perception. *Journal of Empirical Legal Studies, 4*(3), 465–505. https://doi.org/10.1111/j.1740-1461.2007.00097.x

Kéfi, S., Guttal, V., Brock, W. A., Carpenter, S. R., Ellison, A. M., Livina, V. N., Seekell, D. A., Scheffer, M., van Nes, E. H., & Dakos, V. (2014). Early warning signals of ecological transitions:

Methods for spatial patterns. *PLoS ONE, 9*(3), e92097. https://doi.org/10.1371/journal.pone.009 2097

Leach, M., Raworth, K., & Rockström, J. (2013). Between social and planetary boundaries: Navigating pathways in the safe and just space for humanity. In *World Social Science Report 2013* (pp. 84–89). https://doi.org/10.1787/9789264203419-10-en

Leach, M., Scoones, I., & Stirling, A. (2010). Governing epidemics in an age of complexity: Narratives, politics and pathways to sustainability. *Global Environmental Change, 20*(3), 369–377. https://doi.org/10.1016/j.gloenvcha.2009.11.008

Lenton, T. M., Held, H., Kriegler, E., Hall, J. W., Lucht, W., Rahmstorf, S., & Schellnhuber, H. J. (2008). Tipping elements in the Earth's climate system. *Proceedings of the National Academy of Sciences, 105*(6), 1786–1793. https://doi.org/10.1073/pnas.0705414105

Lucas, K., Renn, O., Jaeger, C., & Yang, S. (2018). Systemic Risks: A homomorphic approach on the basis of complexity science. *International Journal of Disaster Risk Science, 9*(3), 292–305. https://doi.org/10.1007/s13753-018-0185-6

Maggs, D., & Chabay, I. (2022). The algebra of the protagonist: Sustainability, normativity, and storytelling. *Innovation: The European Journal of Social Science Research*, 1–12. https://doi.org/10.1080/13511610.2022.2062304

Morgan, M. S., & Wise, M. N. (2017). Narrative science and narrative knowing. Introduction to special issue on narrative science. *Studies in History and Philosophy of Science Part A, 62*, 1–5. https://doi.org/10.1016/j.shpsa.2017.03.005

Okada, N. (2021). Build back better, even before disaster—adaptive design of communicative process, place and practice. In M. Chatterji & P. Gangopadhyay (Eds.), *Contributions to conflict management, peace economics and development* (pp. 27–38). Emerald Publishing Limited. https://doi.org/10.1108/S1572-832320210000029003

Olsson, P., Galaz, V., & Boonstra, W. J. (2014). Sustainability transformations: A resilience perspective. *Ecology and Society, 19*(4). https://doi.org/10.5751/ES-06799-190401

Redman, C. L. (2014). Should sustainability and resilience be combined or remain distinct pursuits? *Ecology and Society, 19*(2). https://doi.org/10.5751/ES-06390-190237

Renn, O. (1999). A model for an analytic—deliberative process in risk management. *Environmental Science & Technology, 33*(18), 3049–3055. https://doi.org/10.1021/es981283m

Renn, O., Burns, W. J., Kasperson, J. X., Kasperson, R. E., & Slovic, P. (1992). The social amplification of risk: Theoretical foundations and empirical applications. *Journal of Social Issues, 48*(4), 137–160. https://doi.org/10.1111/j.1540-4560.1992.tb01949.x

Renn, O., Laubichler, M., Lucas, K., Kröger, W., Schanze, J., Scholz, R. W., & Schweizer, P. (2020). Systemic risks from different perspectives. *Risk Analysis*, risa.13657. https://doi.org/10.1111/risa.13657

Renn, O., Lucas, K., Haas, A., & Jaeger, C. (2019). Things are different today: The challenge of global systemic risks. *Journal of Risk Research, 22*(4), 401–415. https://doi.org/10.1080/13669877.2017.1409252

Rockström, J., Steffen, W., Noone, K., Lambin, E., Lenton, T. M., Scheffer, M., Folke, C., Schellnhuber, H. J., Wit, C. a D., Hughes, T., Leeuw, S. van der, Rodhe, H., Snyder, P. K., Costanza, R., Svedin, U., Falkenmark, M., Karlberg, L., Corell, R. W., Fabry, V. J., … Foley, J. (2009). Planetary boundaries: Exploring the safe operating space for humanity. *Ecology and Society, 14*(2), 32.

Saunders, W. S. A., & Becker, J. S. (2015). A discussion of resilience and sustainability: Land use planning recovery from the Canterbury earthquake sequence, New Zealand. *International Journal of Disaster Risk Reduction, 14*, 73–81. https://doi.org/10.1016/j.ijdrr.2015.01.013

Schweizer, P., Goble, R., & Renn, O. (2021). Social perception of systemic risks. *Risk Analysis*, risa.13831. https://doi.org/10.1111/risa.13831

Shimizu, M., & Clark, A. L. (2019). Nexus of resilience and public policy in a modern risk society. In *Nexus of Resilience and Public Policy in a Modern Risk Society*. Springer Singapore. https://doi.org/10.1007/978-981-10-7362-5

Steffen, W., Richardson, K., Rockstrom, J., Cornell, S. E., Fetzer, I., Bennett, E. M., Biggs, R., Carpenter, S. R., de Vries, W., de Wit, C. A., Folke, C., Gerten, D., Heinke, J., Mace, G. M., Persson, L. M., Ramanathan, V., Reyers, B., & Sorlin, S. (2015). Planetary boundaries: Guiding human development on a changing planet. *Science, 347*(6223), 1259855–1259855. https://doi.org/10.1126/science.1259855

Strogatz, S. H. (2014). *Nonlinear dynamics and chaos: Overview*. With Applications to Physics, Biology, Chemistry, and Engineering. Westview Press.

Swart, R. J., Raskin, P., & Robinson, J. (2004). The problem of the future: Sustainability science and scenario analysis. *Global Environmental Change, 14*(2), 137–146. https://doi.org/10.1016/j.gloenvcha.2003.10.002

Towers, S., Kajitani, Y., Chabay, I., & Okada, N. (2022). *Narratives in the wake of disaster: Comparing the 2018 Hokkaido and 2021 Texas power grid failures*. Preprint at SSRN-id4076708.pdf.

United Nations. (2015). Transforming our world: The 2030 Agenda for sustainable development United Nations A/RES/70/1. In *United Nations* (pp. 1–35).

van der Leeuw, S. (2019). The role of narratives in human-environmental relations: An essay on elaborating win-win solutions to climate change and sustainability. *Climatic Change*. https://doi.org/10.1007/s10584-019-02403-y

Veland, S., Scoville, M., Gram-, I., Schorre, A., El Khoury, A., Nordbø, M., Lynch, A., Hochachka, G., & Bjørkan, M. (2018). Narrative matters for sustainability: The transformative role of storytelling in realizing 1.5°C futures. *Current Opinion in Environmental Sustainability, 31*, 41–47. https://doi.org/10.1016/j.cosust.2017.12.005

World Commission on Environment and Development (UN). (1987). Brundtland report: Our common future. In *Report of the World Commission on Environment and Development: Our Common Future*.

Ilan Chabay Ilan Chabay is Head of Strategic Science Initiatives and Senior Investigator in the Real Deal European Commission project at the Institute for Advanced Sustainability Studies (IASS) in Potsdam Germany as well as Adjunct Professor in the School of Sustainability, Arizona State University. He co-leads KLASICA (Knowledge, Learning, and Societal Change Alliance), an international hub for catalyzing innovative thinking, research, and practice for collective behavior change to sustainable futures with justice and equity. After his first career in innovative laser chemical physics research, he became associate director of The Exploratorium Science Museum (San Francisco), then president of a company designing and producing interactive exhibitions for 230 museums around the world, including Disney, the Smithsonian, and NASA. In 2006 he began his third career in sociology and sustainability science, focusing on understanding the narratives of vision and identity in communities and how they influence collective behavior change toward sustainable futures. In addition to research with 90+ publications in both natural and social sciences, he continues designing games to inspire people for a more sustainable future.

Chapter 4
Deep Uncertainty: Role of Co-Knowledge Production

Mika Shimizu

Abstract This Chapter addresses different layers of uncertainties or deep uncertainty in systemic challenges in and around SDGs. Especially as risks are becoming more systemic, the systemic risks exhibit much more uncertainties, which is a common challenge in implementing SDGs. The critical challenge is that because of "uncertainties" around SDGs, policy-makers or decision-makers tend to delay actions or decision-making which may likely be a major barrier to implementing SDGs. To seek for how to overcome deep uncertainty, this Chapter provides the analysis of structure of deep uncertainty and its impacts on human society with case studies through perspectives of a resilience approach. Based on this analysis, to overcome deep uncertainty, the Chapter suggests paradigm shifts from linear communication and learning approaches to non-linear communication and co-learning-based approaches toward co-knowledge production based on the resilience approach. This kind of shifts or transformation cannot be done only at the macro level but needs collaboration in linking macro-micro levels.

4.1 Introduction and Background

When risks are cascaded and more systemic, they will entail more uncertainties (as discussed in Chapter 3). Thus, in facing systemic challenges which are driven by systemic risks (i.e., complex interactions of natural, human, and social risks at local through global levels) and involve multiple interacting components, that is, the reality of challenges in and around Sustainable Development Goals (SDGs) (see Chapter 1), it is crucial to take into account how to deal with uncertainties, which is one of the underlying common issues in implementing SDGs.

The typical case is that because of "uncertainties" around climate change, natural disasters (e.g., floods, earthquakes, etc.), or infectious diseases, people tend to *delay relevant actions or avoid decision-making,* which may likely be a major barrier

M. Shimizu (✉)
Kyoto University, Kyoto, Japan
e-mail: shimizu.mika.5a@kyoto-u.ac.jp

to implementing SDGs from 2015 to 2030. Although "uncertainties" tend to be understood as a side issue along with major themes that are specified in the SDGs, the in-depth look at the characteristics of uncertainties that are imbedded in risks and their possible impacts on human society can provide critical clues in terms of dealing with systemic risks and challenges in and around SDGs (SDGs 1 to 16). It also gives us a look into communicating and partnering with different stakeholders, including community people who are the key players in implementing SDGs, which is closely related to SDG 17 (partnership).

First of all, what is the meaning of "uncertainty" in this context? It does *not* mean that something is suspicious or the source of information is unknown. It refers to the reality that there is uncertainty, which is difficult to predict even with the latest science. It also means that there are several layers of uncertainties that are interrelated with different kinds of risks and societies, which cannot be addressed by scientists alone. It is also important to note that uncertainty permeates some or all aspects of the problem, including the valuation of outcomes by stakeholders (Marchau et al., 2019).

Regarding the several layers of uncertainties, they are imbedded in risks, including social and economic ones that are translated through interactions in different social and policy systems from macro to micro levels. They result in several unanticipated connections (linkages) that simultaneously occur on different levels. They are unpredictable and form uncertain impacts and outcomes with global to local ramifications across time and space (Shimizu & Clark, 2019), which will form *deep uncertainty*. These impacts often are cascaded across different sectors and local regions or communities to national (or global) in short, mid, and long terms.

Specifically, deep uncertainty refers to situations where experts do not know or the parties to a decision cannot agree with: (1) relationships among drivers and components which shape the future; (2) functions of a system and their boundaries; and (3) outcomes and their values (Lempert et al., 2003, Marchau et al., 2019). Examples of addressing risks with deep uncertainty include deciding where, when, and how to prepare for the future effects of climate change; epidemics and new or deliberately spread pathogens; protecting ecosystems from irreversible loss; and protecting interdependent social and economic systems from disasters (Cox, 2012).

Under conditions of deep uncertainty, policy-makers often find it in their interest to delay action and wait for new information to emerge (Nair & Howlett, 2016). To avoid delayed actions, decision-making in the context of deep uncertainty requires a paradigm shift that is not based on predictions of the future but aims to prepare and adapt, by monitoring how the future evolves and allowing adaptations over time as knowledge is gained (Marchau et. al., 2019). What kinds of paradigm shift is required? How can the paradigm shift be operational? This Chapter addresses these major questions. In doing so, it is essential to take into account uncertainties that are drawn from systemic challenges (see Chapter 1), especially from global to local and vice versa, those which cannot be addressed separately but interact in complex forms, and the different actors including decision-makers, governments, scientists, communities, vulnerable people and their relevant systems or subsystems under uncertainties.

As a clue in seeking answers to the above question, it must be noted that the monitoring and adaptation or adaptive process is closely relevant to the resilience approach which was discussed in Chapter 2. For the validity of adopting resilience in addressing problems under uncertainties, the ecologist Holling (1973) suggested that as future events will be unexpected, we need a qualitative capacity to devise systems that can absorb and accommodate future events, which is relevant to capacity building through the resilience approach. Similarly in the risk management domain, Klinke and Renn (2002) pointed out that, if the uncertainty cannot be reduced by additional knowledge, risk managers need to rely on resilience as the guiding principle for action, that is, designing resilient measures including constant monitoring and investments in diversity and flexibility. This Chapter has built upon these understandings which apply the resilience approach to addressing uncertainties in and around SDGs.

Based on the above, in seeking the major questions, it is critical to look at the details of complex layers of deep uncertainty that are interrelated with complex risks, systemic challenges, and their social impacts. The below provides the following: (a) the structure of complex layers of deep uncertainty, and (b) deep insight into how it may affect or be related to human society and the implementation of SDGs to see the association of deep uncertainty with the resilience approach. Based on this understanding, the author has designed a series of workshops to link some issues that are imbedded in SDGs to local stakeholders and community people to assess how resilience approach-based workshops can influence people's perceptions of uncertainties or drive new insights to address them. As such, the following section draws (c) a snapshot of the design, results analysis, and outcome to assess how the resilience approach can be valid in addressing uncertainties. Furthermore, to extend the application to an another-SDGs related issue, that is COVID-19 and uncertainties, the author conducted the relevant questionnaire survey to the public in 2021. The questionnaire result demonstrates (d) a sign of the need for actions based on the resilience approach, specifically co-knowledge production, which is one of its cores.

Although deep uncertainty is much discussed in terms of decision-makers and policy-making, it is critical to take into account the different layers of societies, especially the overarching scale from global through local or communities, given that the SDGs cannot be implemented without them. Moreover, the systemic dynamics of communities can be a driver for challenging systemic challenges (see Chapter 1) and risks (see Chapter 3). In turn, grasping uncertainties from the overarching global to local or community view will also provide important inputs to decision or policy-makers.

4.2 Layers of Uncertainties

This section articulates the layers of uncertainties that are related to risks in systemic challenges in and around the implementation of SDGs. First, Fig. 4.1 shows a basic category of uncertainties, focusing on "known" and "unknown" relationships

Fig. 4.1 A Basic
known/unknown category of
uncertainties related to risks
in systemic challenges in and
around SDGs

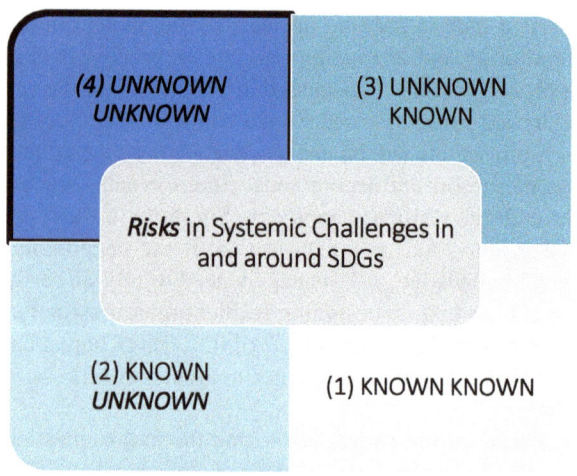

from both expert and public points of view.1 The category includes the following
dimensions:

(1) Known known: What is known by experts is known to the public as well;
(2) Known unknown: What is unknown by the experts is known to the public;
(3) Unknown known: What is known by experts is not known to the public; and
(4) Unknown unknown: Even experts don't know what will happen.

Based on this category, there are two things to note. First, the risk communication
by experts or relevant officials in addressing risks in and around SDGs needs to
take into account what is known including not only the (1) known known, (2) known
unknown, but also (3) unknown known, and (4) unknown unknown, which is relevant
to deep uncertainty, to share those with the public.

Second, communicating uncertainties from (1) through (4) can lead to conveying
the overall picture of risk situations, and avoiding confusion among the public which
can be a barrier in actions or decision making for systemic challenges and SDGs.

Based on the understanding of uncertainties from the perspectives of both experts
and the public, when we took at the in-depth structure of uncertainties, Rowe, W.D.
(1993) presented the structure of scientific uncertainties which is multilayered as
below:

(a) Metrical: uncertainty and variability in measurement;
(b) Structural: uncertainty due to complexity, including models and their validation;
(c) Temporal: uncertainty in future and past states; and
(d) Translational: uncertainty in explaining uncertain results (among different
     stakeholders).

How will these multi-layered scientific uncertainties lead to *uncertainties for
society and individuals*? To put it simply, a general possibility is that, as the various

uncertainties of risks are intertwined, relevant information reaches the final recipients through various stakeholders (including governments and the media), which may create complex uncertainties. Questions arising from individuals who face the intertwined uncertainties in both scientific and social terms may include the following questions:

- "What will happen, when, where, and how?"
- "Which information should I trust?"
- "How should I judge and act?"
- "When will the situation be resolved?"
- "Will things get better or worse?"
- "How do I interpret the information?"
- "How should we interpret the information?"

More technically, as mentioned before, scientific uncertainties imbedded in risks are translated through interactions in different systems or subsystems, resulting in several unanticipated connections (linkages) that occur simultaneously on different levels across time and space (Shimizu & Clark, 2019). They can form social uncertainties and impacts by reflecting the structure of science uncertainties, including: (a) metrical; (b) structural; (c) temporal; and (d) translational uncertainties from social terms (see Fig. 4.2). In particular, the following four dimensions of uncertainties can be pointed out from the perspective of individuals and society along with (a) through (d):

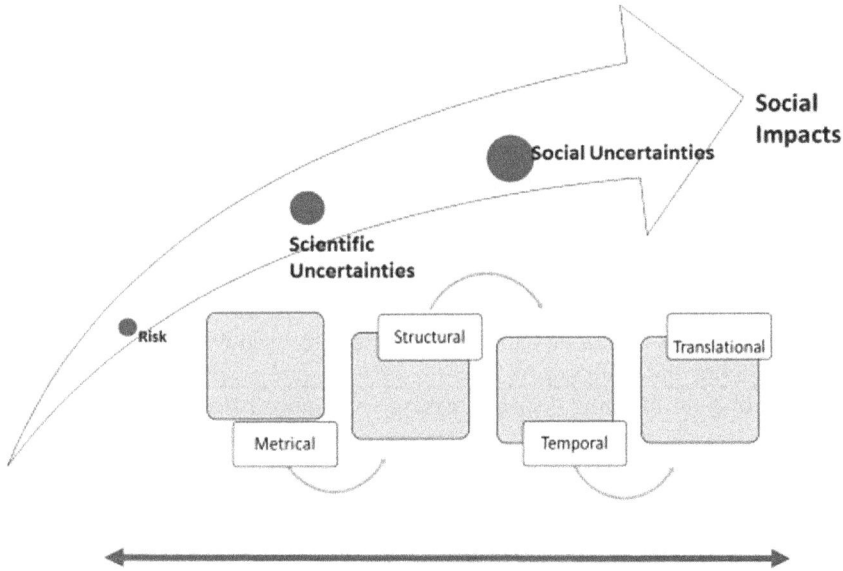

Fig. 4.2   Systemic views from risks through uncertainties to social impacts

- Scientific uncertainty is only one part of a larger problem in considering the overall uncertainties of risks. It is necessary to pay attention to its impact on different contexts of society and individuals through systemic views, especially on vulnerable areas and vulnerable groups, such as the elderly and the disabled.
- Uncertainty with complex dimensions can change significantly across time from the short-, mid-, and long-terms, and across scales through individuals, local to global scales.
- It is extremely difficult to communicate the current state of uncertainties to different stakeholders and the public at one time, partly because the risk is difficult to see and uncertainty is even more difficult and more complex. Therefore, the establishment of a systemic communication process starting from two-way communication among stakeholders is a prerequisite for any action.
- How uncertainty is understood will affect how people face risks and challenges. Based on this, a learning process that encompasses multiple layers and short-, mid-, and long-term perspectives is required, rather than a single temporary learning one.

Taking into account the different kinds of contexts, especially the vulnerable populations and communities, across time and scales, and building systemic communication and learning process is closely related to the resilience approach.

Specifically, such multi-layered uncertainties or deep uncertainty is reflected in the social situations during the novel coronavirus (COVID-19). In particular, the complex situations for more than two years as of 2022 have entailed measurement, structural, temporal, and interpretive uncertainties from both scientific and social points of view, including how and when the virus will mutate, under what circumstances will it develop and in what forms, who will be impacted, what measures should be taken, and when will they be taken. As they are communicated, interpreted, and acted upon through complex social systems or stakeholders (e.g., health care systems, emergency systems, infectious disease tracking systems, vaccine management systems, governments, infectious disease expert committees, hospitals and health care providers, local governments, public health departments, media, and others), uncertainties arise in communication, interpretation, and response. In addition, such a situation especially causes more anxiety among the socially vulnerable, such as the elderly, pregnant women, people with disabilities, and foreigners who do not understand the local language. The impact of uncertainty on individuals and society is also multilayered. This situation makes it difficult to reduce uncertainties, even with established information gathering and analysis. As a result, decision-makers often find themselves delaying actions or avoiding decision-making in advance before serious situations emerge.

4.3 Social Impacts and SDGs

Thinking about the overall social impacts or implications from the above multi-layered uncertainties or deep uncertainty, the following four overarching impacts or implications can be drawn especially in relation to SDGs:

First, as decision-makers delay actions or avoid decision-making, which may result in collective delayed actions or avoidance of decision-making as a whole society because these are not one-time issues and they permeate systemically (i.e., different layers of the problem and stakeholders across time and scale). This can especially happen if there is a lack of alternative measures to turn the negative aspect of deep uncertainty into the positive one (i.e., there is a lack of systemic communication process and co-learning/co-knowledge production process as discussed in the resilience approach in Chapter 2).

Second, it is important to recognize that systemic risks and challenges with such multi-layered uncertainties that are driven by delayed actions or avoidance of decision-making will continuously affect human and social dimensions, especially *vulnerable communities and populations.* It must be noted that the impacts of disasters or climate change do not affect people equally, but largely impact vulnerable communities. For example, the large scale and complex disasters, the Tohoku Disaster (2011) in Japan and Hurricane Sandy (2012) in the United States demonstrated that they had a more significant impact on vulnerable populations, including the aged people, disabled people, children, and people in poverty or vulnerable communities (Shimizu & Clark, 2019). Moreover, the impacts of climate change are unduly experienced by vulnerable and marginalized populations. The inter-linkage among these issues was identified in the 2012 Intergovernmental Panel on Climate Change (IPCC) report, which articulates how exposure and vulnerability to weather and climate events determine disaster risks. Based on the known knowledge, facing complex uncertainties or deep uncertainty from systemic views makes it necessary to pay attention to the impacts of multi-layered or deep uncertainty, especially on vulnerable people and communities.

Concerning vulnerable populations and communities, SDGs include the term "vulnerable" in separate targets in particular goals, such as Goal 1 (no poverty), Goal 2 (no hunger), Goal 4 (quality education), Goal 6 (quality water and sanitation), and Goal 11 (sustainable cities and communities). For example, within Goal 11, target 11.2 states that "By 2030, provide access to safe, affordable, accessible and sustainable transport systems for all, improving road safety, notably by expanding public transport, with special attention to the needs of those in vulnerable situations, women, children, persons with disabilities and older persons." The vulnerability incorporated here may be related to land, poverty, water, education, climate change, health, and others, which are linked in systemic ways. In other words, the relevant problems do not exist in a dot but are interconnected and influence each other in complex ways.

Given the above systemic nature of vulnerabilities and specific issues in SDGs, delaying actions or avoiding decision-making due to complex uncertainties or deep uncertainty may lead to more impacts on the vulnerable communities and population. This, in turn, will largely impact on the implementation of SDGs (see Fig. 4.3) as the lack of a systemic approach creates the situation of "missing links." Although grasping the systemic nature of the challenges is key in addressing the specific problem in the specific goal, the systemic perspectives are hard to grasp by just looking at the specific goal and target. The kind of dot-based approach, which

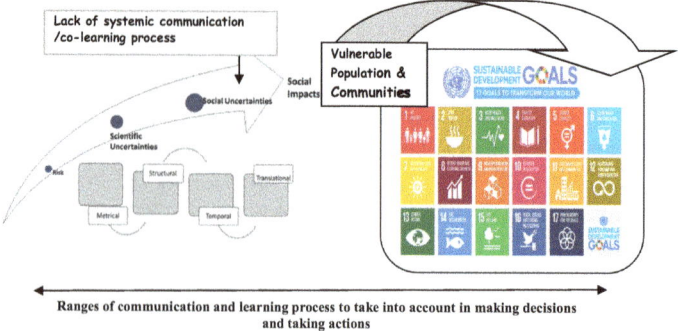

Fig. 4.3 Social impacts and SDGs

exists everywhere, does not necessarily pave the pathway for addressing systemic challenges in and around SDGs.

As such, given the systemic nature of risks and challenges in and around SDGs, it is critical to spotlight the systemic approach in addressing systemic risks and challenges in and around SDGs with multi-layered uncertainties or deep uncertainty. This aspect is linked to the necessity of adopting the resilience approach.

Third, the vulnerable population does not necessarily refer to passive people who just receive goods and funds from governments or aid groups. They are sometimes active innovators or change-makers. This has been proved by past disasters or the ongoing local or community practices for systemic challenges. That kind of innovation or change-making can emerge only when people and communities engage in co-knowledge production with different stakeholders inside and outside the relevant communities (case studies are shown in Chapters 5 and 6, and see, for example, Shimizu & Clark, 2019).

Fourth, based on the above themes, especially the third point, to manage uncertainties or deep uncertainty from the SDGs point of view by incorporating a part of the resilience approach in a systemic process, it is critical to incorporate the followings into the management process:

- Open information and data, and their integration; reviewing those through the co-production knowledge formation process;
- Systematizing schemes for scientists to locate in the policy formation process;
- Clearly and regularly framing risks and uncertainty based on the above schemes with different stakeholders through interactive communication; and
- Operationalizing collaborative learning, and knowledge-building schemes among scientists, policy communities, and other stakeholders including different communities.

4.4 Praxis of the Resilience Approach and Co-Knowledge Production: Global to Local

4.4.1 Global to Local

Based on the above understandings of multi-layered uncertainties or deep uncertainty in association with vulnerabilities, communication, and co-knowledge production, the author designed a series of local workshops to link some issues that are imbedded in SDGs to local stakeholders and communities. This was done to assess how the resilience approach-based workshops can change their perceptions of uncertainties and drive stakeholders' insights toward actions by taking the case of the possible Nankai Traph earthquake in Japan, which could result in other complex and large-scale uncertainties. Specifically, the Nankai Traph has the characteristic of "deep uncertainty," since even the best scientists cannot predict when, where, and how it will occur in deterministic terms (Hashimoto, 2020). The following draws a snapshot of the design, results analysis, and outcome to assess how the resilience approach can validly address uncertainties. Specifically, the challenges which could occur during the Nankai Traph earthquake are interrelated with communication, co-knowledge production, vulnerable population, and communities.

4.4.2 Design of Co-Knowledge Production Based on the Resilience Approach

The series of workshops was designed based on the resilience approach, by incorporating the following (1)-(5) components in the workshops, to explore how to manage uncertainties that are related to the Nankai Trough earthquake in Japan, in collaboration with research partners including natural, social, and humanities scientists and practitioners. Specifically, the series of workshops were organized from 2016 through 2018 under the theme of "earthquake risk and uncertainty." It assumes a major earthquake along the Nankai Trough may occur in near future, targeting different stakeholders, including school teachers, businesses, local or national governments, voluntary disaster management organizations, and the public.

(1) **Linkage between emergency and daily life:** In an actual emergency setting, participants experience first-hand disruption that could occur during the emergency and reflecte on their daily risk communication.

(2) **Scenario-based:** Participants go through the scenarios created by the Cabinet Office's "Subcommittee on the Predictability of Large-Scale Earthquakes along the Nankai Trough," which are divided into several stages with detailed assumptions including season, timing, location, surrounding conditions, and how the situations could change. Based on this, participants are guided to identify possible actions by taking into account not only earthquake scenarios, but also natural and social

environmental conditions, and situational scenarios of individuals, their communities and their organizations.

(3) Field-based: Instead of simply providing knowledge, the program focuses on what could happen in the field, taking into account the situations and specific contexts of each stakeholder.

(4) Dialogue-based: It is important to promote a dialogue environment among different stakeholders (e.g., experts, citizens, government and business, etc.) by facilitating dialogues between through interactive communication.

(5) Creation of collaborative knowledge: Participants' awareness and insights are drawn out through collaborative work. Collaborative knowledge is created through dialogues, recording, and sharing knowledge.

In particular, the program was constructed with particular emphasis on ① and ② in Fig. 4.4. Especially for the former one, the following programs were incorporated:

- Dialogue between scientists (including scientific explanations) and practitioners; and
- Collaborative work based on scenarios.

In the "collaborative work," groups of several people were formed (composed of the same or different stakeholders in a group, depending on the situation), assuming the real situation and emphasizing diversity.

Operationally, based on the premise of "going beyond the boundaries of each stakeholder, building a relationship of trust through communication to overcome pain and discomfort when encountering problems, practicing together based on the actual situation, reviewing the current situation, and learning together," the following elements were incorporated into the workshop:

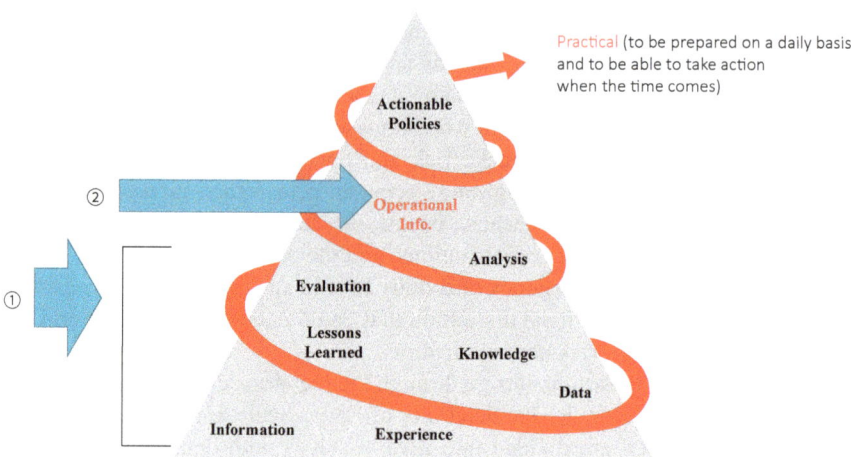

Fig. 4.4 Design of co-production knowledge based on the resilience approach (Shimizu & Clark, 2019)

- To provide (a) science in practice, (b) implications of "uncertainty" from both scientific and social perspectives, (c) possible effects on our daily lives and social functions, and (d) scenarios that are relevant to the field, provided through dialogue methods and two-way communication along with each stakeholder's perspective.
- To present the questions to the participants on a "step-by-step" process (Note: depending on the scenario, various situations could be assumed for the short-, mid-, and longterms, especially for the Nankai Trough earthquake), based on the assumption of daily life.
- To draw insights through constructing dialogues, making notes, and sharing the results continuously and consistently at each stage, and extract collaborative knowledge from the aggregate of these processes.

Furthermore, in step ②, a session to create collaborative knowledge was established, where the results of step ①were consolidated and systematized, keys for the next actions were extracted, and a flow was created for all participants to share issues and possibilities.

In this series of steps, based on Fig. 4.5, special attention was paid to: (i) what kind of results would be produced by the dialogue through collaborative work "based on scenarios" within and between communities (groups); and (ii) what kinds of directions could be identified among all participants by accumulating and systematizing those results (scaling up through accumulation).

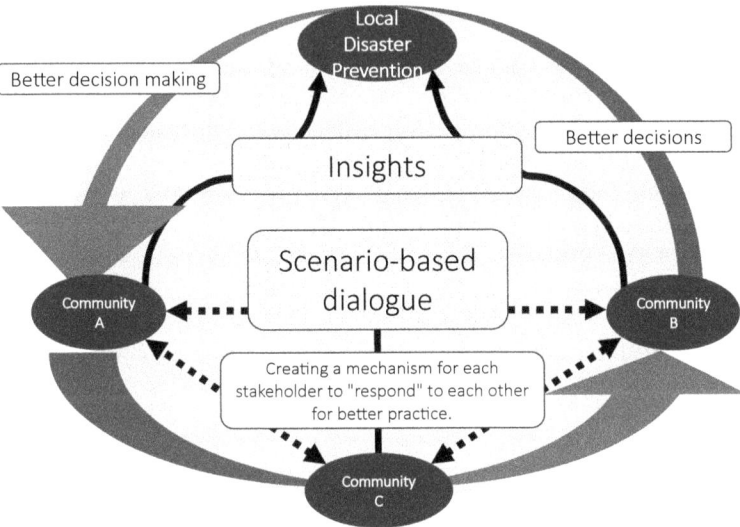

Fig. 4.5 Design of co-production knowledge based on the resilience approach from micro views (Shimizu & Clark, 2019)

4.4.3   Results

As an indication of the results of the series of workshops designed based on the resilience approach, a summary of participants' insights that were elicited during the series of workshops, and a summary of what the participants were asked to write after the workshops when they returned to their daily lives to reflect on what they experienced during the workshops (reflections) are shown in Boxes 1 and 2:

Box 1: From the Participants' Insights

- In risk communication that is related to uncertainty, an "intermediary" person (middleman) is needed to mediate between scientists ⇔ government ⇔ citizens.
- Even if the information is uncertain, it is important to train ourselves to think of it as a reality, not in a vacuum.
- If we have even a little basis for judgment, we can think about uncertainty through several processes and understand it. I have been used to acting only after something happens, but now I recognize the importance of thinking and acting from the stage when information is uncertain.
- Given uncertain situations, I realized that it is necessary to take into account the vulnerable and handicapped people in their current preparedness for disasters.
- Through the workshop, questions which never came up in my mind were raised, such as: Can evacuation centers be operated for days, weeks, or even months of evacuation life? Will the budget for the shelters be sufficient?
- Given the scenarios under uncertain conditions, three days is not enough to stockpile foods…seven days? What is the emergency response under emergency conditions with the uncertainties?
- I realized through my own experience in the workshop that I am not prepared for the disaster and more than that, it is even more difficult to deal with uncertainty.
- Uncertainty is not only about general events, but also about what will happen to us.

Box 2: From Participants' Reflection

- We need to shift from "informing" people about earthquake risk to "working together to identify issues around the earthquake." In such efforts, it is necessary to think together about how to respond to "uncertain information" when things do not happen as expected.

- During the workshop, I was able to think about earthquakes and uncertainty several times with different stakeholders through specific simulations. This allowed me to gain a variety of "insights" that do not come to mind on a daily basis. If I consider this alone, no matter how hard I tried, I would imagine things from my own perspective. By discussing and examining things with people in different positions, I was able to get closer to seeing and thinking things objectively, even if only a little.
- The kind of workshop can be considered as one of the "evacuation drills" to discuss "uncertain information about earthquakes" with family, community, school, and workplace.

4.4.4 Assessment

The results indicated that in addition to the day-to-day disaster reduction actions, specific actionable views were drawn out as to what considerations and further efforts are needed in light of the reality of layers of uncertainties or deep uncertainty. Given these results, it is clear that the co-knowledge creation setting based on the resilience approach is effective to some extent in dealing with complex risks and uncertainties.

4.5 Needs for Co-Knowledge Production During Covid-19

How does the public understand complex uncertainties or deep uncertainty? In the middle of COVID-19 (2021), the author has conducted a relevant questionnaire survey of the public in Japan. This section introduces a part of the survey results and assesses them using the resilience approach to see how they understand and adapt to complex uncertainties or deep uncertainty.

The questionnaire survey was conducted among the general public from March 15 to 28, 2021, under the major theme: "How did you perceive and deal with uncertainty during the Covid 19?" The target population was 18 years old or older nationwide. A random online survey (randomly notified about 14,000 people, mainly social networking sites users) was conducted, and 352 people responded.

Major survey questions and results 1 through 4 are shown below. Based on the 1–4, overall, four major findings are drawn: First, although the majority of the responders think they get adequate information about COVID-19 (well informed 58, and informed 244) as seen from (1), the majority think that the way they receive information or communication has problems (38% think that information is difficult to understand; 30% think a two-way communication is needed) as seen from (2).

(1) **Survey Question and Result 1**

From your personal perspective: Did you get adequate informations about
Covid-19 pandemic?

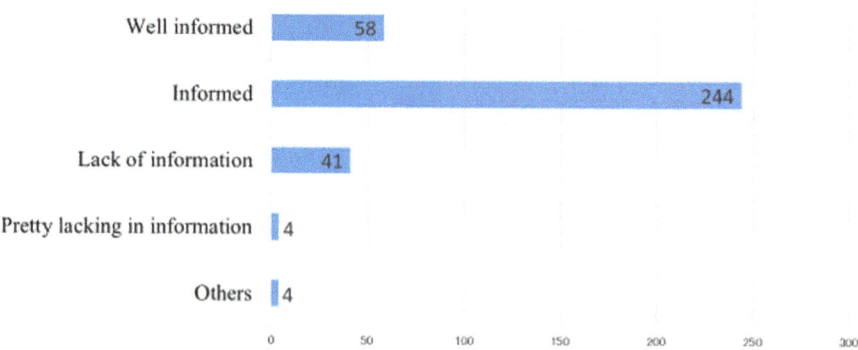

(2) **Survey Question and Result 2**

From your personal perspective: How do you think about Covid-19 informations
provided by public institutions? (Multiple answers allowed)

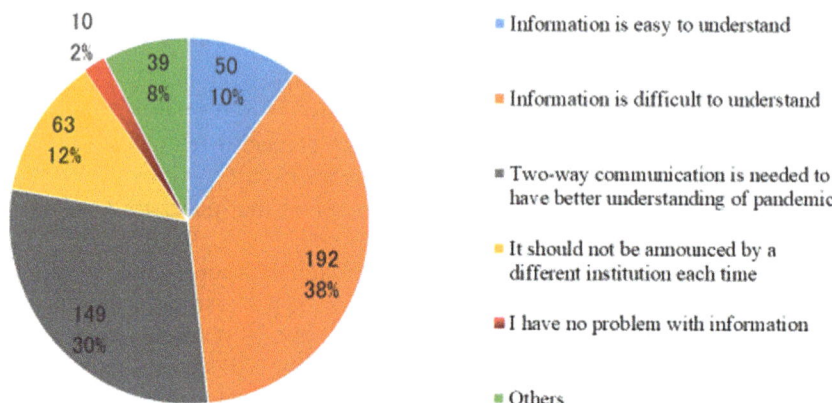

Second, in terms of uncertainties that people face, as seen from (3), in addition
to factors that are directly related to COVID-19, it is found that there are different
sources of uncertainties that are mainly related to social and economic ones.

Third, responders tend to think that such anxieties and uncertainties can be reduced
by acquiring information and knowledge, while "improvement through communication by experts and the government" was cited by about half of the total respondents.

On the other hand, about 60% mentioned not only receiving information but also learning on their own initiative is crtical. This percentage was noticeably higher than any other responses. On the other hand, only about 10% answered "there is nothing I can do about it."

(3) **Survey Question and Result 3**

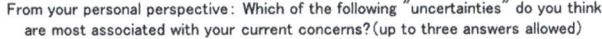

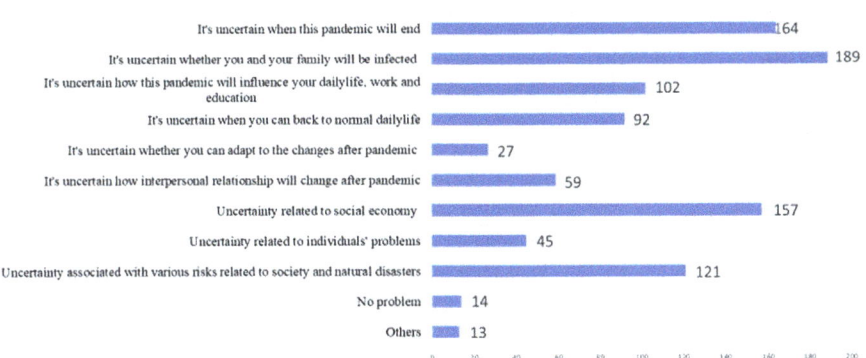

(4) **Survey Question and Result 4**

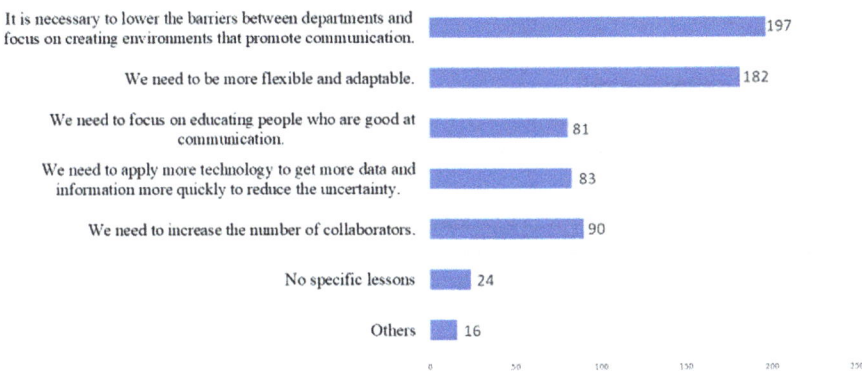

Fourth, as seen from (4), regarding the lessons learned from COVID-19 from the community, team and organizational perspectives, more than 50% of the respondents cited "the need to lower the barriers between organizations and create a place and environment that encourages communication" and "improving the flexibility and adaptability of individuals" (197 and 182 respoendents, respectively).

Furthermore, when asked an open-ended question about what they think is important in times of uncertainty amidst COVID-19, and what they would like to do in collaboration with different stakeholders, the main examples of responses are shown in Box 3.

BOX 3: Responses to the question, "What do you think is important in times of uncertainty, and what would you like to do in collaboration with different stakeholders?"

- Redefining the community and finding ways to communicate in the community more deeply online.
- Since it is important to develop the habit of in-depth thinking for all of us, it is necessary to improve or develop educational systems (courses, teaching methods, etc.) for this purpose.
- It is a necessity to develop the ability to judge whether the information given to citizens is sufficient and correct.
- I believe that it is necessary to improve literacy on understanding uncertainty related to science.
- Since we are already living in uncertain times which is not limited to COVID-19, it gave us an opportunity to think about the uncertain age, which can be a chance to think about what is important for us once again to create better social values.
- It is necessary to have an interactive environment where people can teach and learn from each other in a frank manner, not just receive information and knowledge.
- Connectivity with different kinds of people, that is, not just people close to myself but also outside my own preference or community.
- I feel it is necessary to develop the ability to relate to others from an early age so that we can build relationships that help each other without isolating ourselves.
- Creating a space for dialogue.
- Creating a place of experience (trials, and learning process) to generate the belief and action of being a collaborative partner, rather than building walls around allies or enemies.
- Self-affirmation and self-determination, and education that fosters them. Decision-making methods that do not follow majority rules. Persistence in dialogue and mindsets that lead not to division into friend and foe.

These kinds of responses indicate that there is a pronounced tendency to try to overcome difficulties brought about by COVID-19 by adapting through collaboration or starting a new trial based on improving the way of communication and collaboration. For this, people tend to think that it is necessary to review the traditional "community" and create a place and environment for in-depth thinking and dialogue with different people, rather than the traditional learning methods (e.g., lectures, etc.).

In terms of how the survey results are associated with the resilience approach, the trends drawn from the questionnaire are, overall, highly compatible with the approach, even though the respondents were not informed about it or the relevant ways of thinking or approach before answering the questionnaire. In particular, in light of the co-learning and co-knowledge production process which is a leverage point of the resilience approach, there is an awareness that the one-way, non-interactive communication by experts and public agencies hinders the appropriate transfer of information, and does not address the anxiety and uncertainty of many people who believe that they are adequately or somewhat informed. As such, there is a need for interactive communication with a co-learning process to connect citizens with policies. Moreover, the need to build or restructure co-knowledge and collaborative knowledge production processes and create environments to enable the process is articulated in the survey.

4.6 Conclusion

Reflecting on the questions that are specified at the beginning of this Chapter (i.e., what kinds of a paradigm shift is required and how the paradigm shift can be operational), the following points can be drawn by monitoring complex uncertainties or deep uncertainty in systemic challenges in and around SDGs and some case studies.

First, a paradigm shift from dots-based separate approaches **to linkage-based approaches** (c.f. linking macro–micro levels); and **from linear communication and learning approaches to non-linear communication and co-learning-based approaches** is necessary. In other words, since complex uncertainties or deep uncertainty from systemic challenges (see Chapter 1) cannot be addressed in linear ways such as predictions → policy measures → local actions, non-linear communication and co-learning-based approaches are crucial with the triple-loop learning processes (Chapter 2) by linking decision-makers, governments, scientists, communities, vulnerable people and their relevant systems/subsystems under uncertainties.

Second, built upon the first point, a paradigm shift from separate approaches at global and local levels **to global to local (or vice versa) overarching approaches** is a necessity toward **co-knowledge production.** This means it is important to create channels or schemes that enables different kinds of knowledge not only academic or expert knowledge beyond disciplines but also local and community knowledge to synergize. Through co-knowledge production processes, capacity building to adapt to systemic challenges can be possible for different stakeholders including the public, so that they can engage in relevant actions given the systemic challenges.

Third, especially in terms of how the paradigm shift can be operational, it is critical to operationalize co-knowledge production paying attention to local, community or human contexts. Although some major players at the global level pay attention to community activities all over the world and try to collect the best practices of different communities related to SDGs, it is not enough in that they tend to be addressed as separate case studies with less attention to different contexts. Those best practices need to be grasped both in details and the whole and in relations with local through global perspectives, which need to be shared different stakeholders for co-knowledge production. Furthermore, while local and community activities which do not necessarily specify "SDGs-related" activities often have linkage-based best practices and efforts that have been nurtured through local or community people, this kind of knowledge can greatly contribute to solving problems in and around SDG implementation, especially if their knowledge interacts with other kinds of knowledge including expert and policy knowledge to scale up problem-solving oriented actions.

In a nutshell, the above points can converge into the necessity of updating a "co-knowledge production system" through incorporating components of the resilience approach into the system, which is a leverage point to enabling resilience or operationalizing the resilience approach.

For many decision-makers or stakeholders, addressing deep uncertainty is tricky. However, it must be noted that deep uncertainty is, in most cases, a given in and around SDGs. Whether to consider them in making a decision is a choice. Ignoring deep uncertainty looks attractive but a decision that ignores it ignores reality (Marchau et. al., 2019).

References

Boin, A., Lodge, M., & Luesink, M. (2020). Learning from the COVID-19 crisis: An initial analysis of national responses. *Policy Design and Practice, 3*(3), 189–204.

Cox, L. A., Jr. (2012). Confronting deep uncertainties in risk analysis. *Risk Analysis: An International Journal, 32*(10), 1607–1629.

Hallegatte, S., Shah, A., Brown, C., Lempert, R., & Gill, S. (2012). Investment decision making under deep uncertainty--application to climate change. *World Bank policy research Working Paper.*

Hashimoto M. (2020). Earthquake science in making and the Nankai trough temporary information on Nankai trough earthquakes. *Journal of Japan Society for Natural Disaster Science, 39*-1, 5–9. (in Japanese)

Holling, C. S. (1973). Resilience and stability of ecological systems. *Annual Review of Ecology and Systematics, 4*, 1–23.

Intergovernmental Panel on Climate Change. (2012). Managing the risks of extreme events and disasters to advance climate change adaptation. Retrieved from https://www.ipcc.ch/pdf/special-reports/srex/SREX_Full_Report.pdf. Accessed February 25, 2022.

Klinke, A., & Renn, O. (2002). A new approach to risk evaluation and management: Risk based, precaution based and discourse based strategies. *Risk Analysis: An International Journal, 22*(6), 1071–1094.

Lempert, R. J. (2003). Shaping the next one hundred years: New methods for quantitative, long-term policy analysis. RAND.

Marchau, V. A., Walker, W. E., Bloemen, P. J., & Popper, S. W. (2019). *Decision making under deep uncertainty: from theory to practice.* Springer Nature.

Nair, S., & Howlett, M. (2016). From robustness to resilience: Avoiding policy traps in the long term. *Sustainability Science, 11*(6) Springer.

Rowe, W. D. (1994). Understanding uncertainty. *Risk Analysis, 14*(5), 743–750.

Shimizu, M. and Clark A. (2019). *Nexus of resilience and public policy in a modern risk society*, Springer.

Mika Shimizu Mika Shimizu is an Associate Professor in Graduate School of Advanced Integrated Studies in Human Survivability, Kyoto University. Her long years' experiences as a policy researcher in East-West Center in Washington DC and Honolulu, Hawaii in the United States greatly contributed to publishing this book. She holds an M.A. from American University and a Ph.D. in International Public Policy from Osaka University (2006). She has been extensively involved in interdisciplinary and transdisciplinary research projects related to disasters/infectious diseases, sustainability, and climate change issues with the focus on resilience. Her major publications include Nexus of Resilience and Public Policy in a Modern Risk Society (Co-Author: Allen Clark, Springer, 2019).

Part III
Practical Perspectives

Chapter 5
SMART Governance Under Persistent Disruptive Stressors to Enhance Community's Dynamic Resilience: Case of Chizu Town, Japan

Norio Okada

Abstract This Chapter discusses community resilience from the viewpoint of community's coping capacity challenged for surviving and regrowing towards sustainable future. Such a community challenge is called "rural decline" or more broadly "community decline" to refer also to urban areas. An integrative perspective is introduced to characterize (rural) community decline. Rural decline is caused and accelerated by different natural and social stressors such as loss of population and aging, social and economic changes as well as natural disasters and climate changes. The study consists of two aspects which are mutually interrelated. One is to propose conceptual and methodological frameworks in order to strategically build up dynamic resilience for the community at stake. The second is to make full use of field-based evidence accumulated through the author's three decade-long studies conducted in Chizu, Tottori, Japan. The case history demonstrates that people have increased adaptively their coping capacity through the long years' participatory process. as an adaptive process for SMART community governance under persistent disruptive stressors. It is argued that communicative spaces are critical, especially to enable community's dynamic resilience.

5.1 Introduction

This Chapter highlights rural decline issues in Japan and argues how to strategically enhance community's coping capacity to survive and transform towards sustainable future. Rural decline is caused and accelerated by different natural and social stressors such as loss of population and aging, social and economic changes as well as natural disasters and climate changes. This is a core challenge in addressing the rural decline occurring across Japan.

In parallel, urban areas are also suffering from community decline issues due to seemingly similar persistent stressors. To overcome such community challenges,

N. Okada (✉)
Kwansei Gakuin University, Nishinomiya, Japan
e-mail: kyotookanori@gmail.com

community's dynamic resilience (CDR) has to be strategically enhanced. Here "dynamic resilience" means the community's cohesion and coping capacity to adapt and transform even under Persistent Disruptive Stressors (PDSs) (Okada, 2010, 2016, 2018a). Based on the author's 30-year engagement in the mountainous community Chizu Town, Tottori Prefecture, Japan, a unique participatory approach called "Zero-to-One Movement" (see the following sections for details) has been strategically studied. The study has shown that the community has adaptively increased its coping capacity. The unique participatory process can be presented as an adaptive process for SMART community governance under persistent disruptive stressors— "S" represents small-sized and survivability-minded, "M" modest-scale and multiple-stakeholder involved, "A" anticipatory and adaptive, "R" risk-concerned and responsive, and "T" is transformative. –One of the lessons learned from the Great Hanshin-Awaji Earthquake disaster in 1995 in Japan was a lack of people's awareness and capacity to reduce disaster risks. Hence people recognized the need to address citizen-led participatory approaches to disaster risk reduction before disasters, as well as for disaster recovery and revitalization after disasters (Okada, 2010, 2016). Since then, a series of large-scale, natural hazard-induced disasters have occurred across Japan—such as the 11 March 2011 Tohoku (Eastern Japan) Earthquake disaster, the 14 and 16 April 2016 Kumamoto Earthquake disaster, and most recently the July 2018 Western Japan Heavy Rainfall Disaster, which claimed more than 220 lives and devastated many areas of western Japan (The Asahi Shinbun, 8 July 2018). All confirmed the ever-increasing need for citizen-led participatory approaches. Based on the above, the following sections will provide the conceptual foundation including adaptive process for SMART community governance under combined risks of rural decline and disaster devastation. Then the Case of Chizu Town, Tottori, Japan will be shown to illustrate the preceding conceptual discussions.

5.2   Conceptual Foundation

5.2.1   *Kojo, Kyojo, and Jijo Partnership for Disaster Risk Reduction in Japan*

As explained above, people have been increasingly exposed to disaster risks, particularly since the Great Hanshin-Awaji Earthquake disaster in 1995. To reduce such disaster risks, each community's coping capacity with disasters has to be strategically improved, in parallel with the importance of governments' roles and leadership in disaster risk reduction (Okada, 2016). To understand the emerging need for such a significant change in disaster planning and management in Japan, one must understand the contrasts between *Kyojo* (neighborhood or community self-reliance), *Jijo* (individual or household self-reliance), and *Kojo* (government assistance). Realizing the limitations in the government's capacity to provide immediate and direct support for relief and recovery after a large-scale disaster, Japan has shifted more towards increasing both the *Kyojo* and *Jijo* self-reliance roles to depend less on *Kojo*, which in the past has been the major agent to mitigate disasters. This is a good example of disaster risk governance in public–private-individual partnership. Though some progress has been made to improve community disaster resilience with this partnership approach, it has its own limitations. One problem is how to initiate such a participatory approach and who will be able to facilitate the whole process. Another is how to involve community stakeholders who have not been previously engaged in disaster issues. It is a question of engaging a broader range of stakeholders in social processes (Okada, Fang et al., 2013a, 2013b; Okada, Na et al., 2013a, 2013b; Okada, 2006a, 2006b). To address this question properly, we need a much broader scope by not limiting ourselves to disaster concerns, but by also addressing a wide array of day-to-day community concerns.

To overcome such community challenges, community's dynamic resilience (CDR) has to be strategically enhanced. As mentioned above, "dynamic resilience" which is supposed to enhance the community's cohesion and coping capacity to adapt and transform even under Persistent Disruptive Stressors (PDSs) is a key in addressing the challenges (Okada, 2010, 2016).

5.2.2 Addressing the Whole Mix of Community Problems: The Machizukuri Approach in Japan

Okada and Fang et al., (2013a, 2013b) proposed systematic conceptual models for understanding the *machizukuri* (bottom-up citizen-led participatory approach) in Japan.[1] Figure 5.1 illustrates the multilayer common spaces (an extension of the concept of infrastructure) for a city, region, or neighborhood community as a living body (Okada, 2006b). In the context of this pagoda model, *machizukuri* is more appropriately applied at a neighborhood community scale, rather than at a wider scale, such as city or region. Applied to a neighborhood community in the context of a five-storied pagoda model, it starts with the fifth layer (daily life), followed by the fourth (land use and built environment), and the third (infrastructure). By comparison, *toshikeikaku* (urban planning) focuses mainly on the fourth and third layers. Another point of contrast is that *machizukuri* requires citizen involvement to induce attitudinal or behavioral change, while this issue is not essential for *toshikeikaku*. This kind of *machizukuri* process is a human-centered approach, and has its basis in the fifth layer—people's lives. It is commonly initiated by people at the top and then proceeds to the lower layers, and back and forth. The *machizukuri* process is the opposite of the traditional type of urban planning headed by experts and administrative bodies. By convention we call the latter a top-down approach and the former a bottom-up approach, but the pagoda model (Fig. 1) challenges this mindset and argues that it could be upside-down, depending on our standpoint.

---

[1] For *machizukuri* see Okada (2016).

Fig. 5.1 Five-story pagoda model of the common spaces in a city, region, or neighborhood community

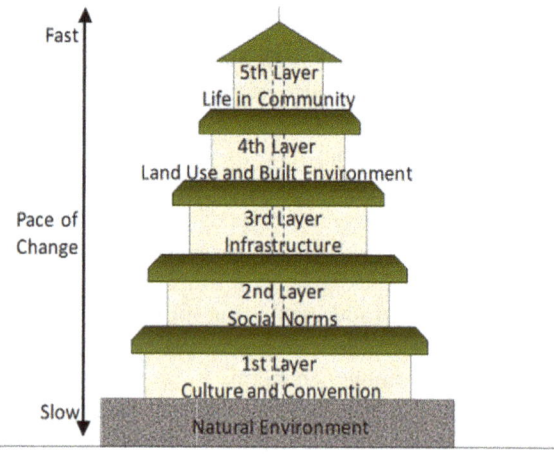

The dynamic processes of *machizukuri* in implementing such a change can be explained and systematically modeled by the nested Plan-Do-Check-Action (PDCA) cyclic structure (Okada, 2006a, Fig. 2). The "plan" segment in each cycle is set as a hypothetically-workable countermeasure for their committed small change. It is designed to evolve step by step, from small through middle to large cycle, thus changing adaptively and spirally. In structure such a PDCA cyclic process is analogous to the methodological approach of adaptive management. As a hypothetical countermeasure, the plan characterizes the essential feature of active adaptive management (see Sect. 3.6).

Later in **4.** and **9.,** this PDCA cyclic process will be further elaborated and incorporated into "SMART community governance" (Fig. 5.2).

Fig. 5.2 Nested structure of the Plan-Do-Check-Action (PDCA) process designed to evolve spirally

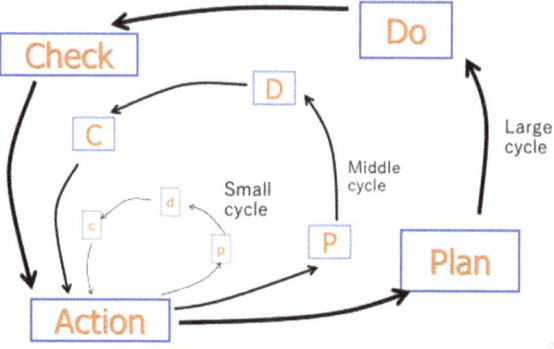

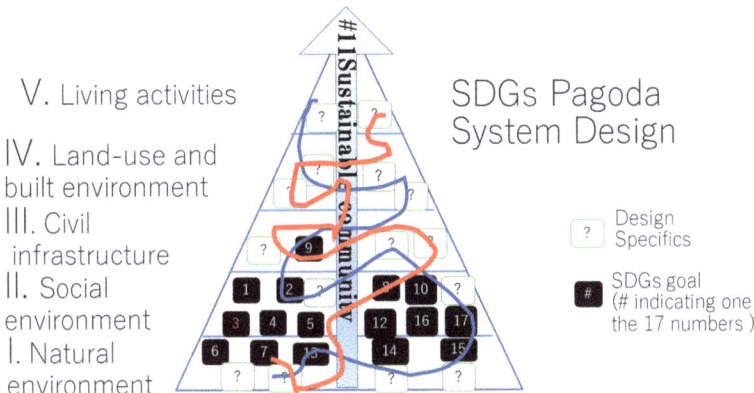

Fig. 5.3 An Example of placing SDGs in five layers of pagoda model

5.2.3 Pagoda Model to Represent the Whole Mix of Community Challenges Towards Sustainable Future

The above pagoda model has a wide range of applicability to address the whole mix of community challenges and can be applied to strategic engagements in holistic features of sustainability issues for a community at stake. Specifically, this model can be applied to the Sustainable Development Goals (SDGs), also known as the Global Goals, adopted by the United Nations in 2015, especially in that all 17 goals in SDGs integrated—they recognize that action in one area will affect outcomes in others, and that development must balance with social, economic and environmental sustainability. Moreover, it should be emphasized that all 17 goals are mutually affecting each other in principle and the essence of sustainability is the whole mix of entire issues. This can be explained by "systemic" nature in and around SDGs in the dictionary sense, which is, "a systemic problem or change is a basic one, experienced by the whole of an organization or a country and not just particular parts of it" (Cambridge Dictionary. Okada, [2021]) (Fig. 5.3).

5.3 Adaptive Process for SMART Community Governance Under Combined Risks of Rural Decline and Disaster Devastation: The Case of Chizu Town, Tottori, Japan

5.3.1 Overview and Case History

This section provides the detailed case history of efforts of a region in Japan, Chizu Town, Tottori, which has gone through adaptive process for the above

Fig. 5.4 Location of Chizu Town, Tottori Prefecture, Japan (http://www1.town. chizu.tottori.jp/dd.aspx?men uid=1)

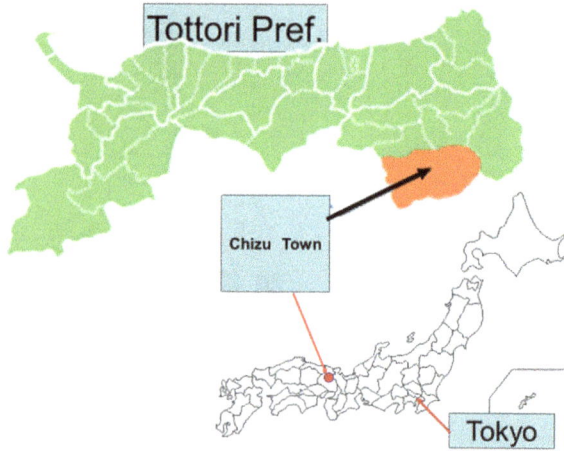

"SMART" community governance under combined risks of rural decline and disaster devastation.

Figure 5.4 shows the mountainous location of Chizu Town, Tottori Prefecture, Japan. This town is a typical type of mountainous municipality in the western part of Japan, with a population of 6,689 as of October 2018. It is located about 500 km in direct distance west of Tokyo. The town is a part of Tottori Prefecture (whose capital city is Tottori City), which faces the Sea of Japan. As of 2018, this prefecture has a population of about 570 thousand, and it ranks at the bottom in population among the 47 prefectures in Japan. Chizu Town is a very old community known to exist since the ninth century but especially since the 1950s until now, it has been suffering from constant loss of population and rural decline. The major industry has been forestry (timber industry) combined with rice and other agricultural production. A high quality of cedar timber has long been known as a local brand in Japan.

Figure 5.5 shows that Chizu Town has long been suffering from loss of population, particularly of younger people under 65, resulting in the gradual increase of the proportion of people over 65. Many other municipalities, particularly in rural areas, have also long been suffering a similar kind of community decline that is commonly called the *Kaso* (hollowing out) syndrome in Japanese. Governmental efforts have been immense to stop this degradation, by constantly supplying subsidies and making public investments in suffering rural areas, but without much success. People in most rural area communities tend to become less able to cope with their problems by themselves. Chizu Town used to be among them until the early 1980s, but turned out to be one of the exceptions at about the time the author was invited to the town by a local leader and his local vitalization team in 1985.

Since then, the author has been conducting community-based field studies in Chizu, in close collaboration with social scientists led by T. Sugiman (2006; see also Okada & Sugiman, 1996; Kawahara & Sugiman, 2003). The field studies are characterized by transdisciplinary research challenges, and strategically designed to be a three decade-long evidence-based investigation and systematic documentation

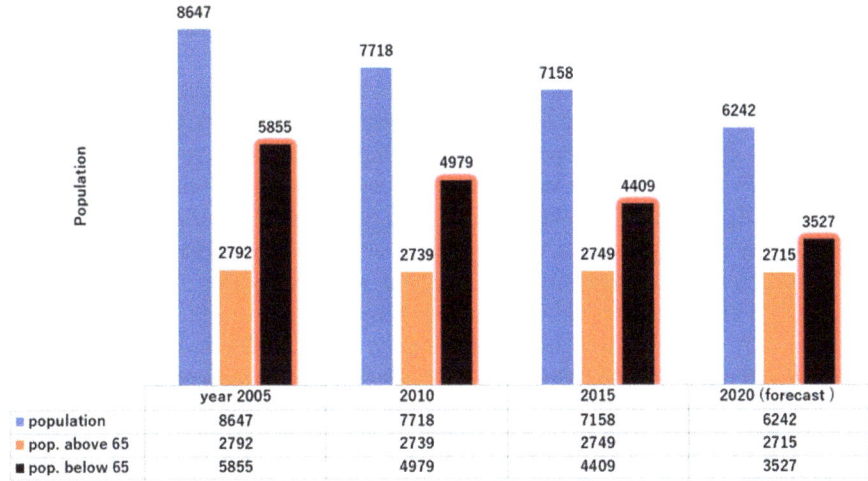

Fig. 5.5 Population change in Chizu Town, Tottori Prefecture, Japan, 2005–2020 (http://www1. town.chizu.tottori.jp/dd.aspx?menuid=1)

of the three decade-long process of Zero to One *machizukuri* (see the box 1 for the history), a unique resident-led participatory approach characterized by the revitalization challenges of their local communities and a step-by-step transformation of their social systems.

Box 1: Three decades of Zero-to-One *Machizukuri* transformation processes in Chizu

Decade I: 1987–1997. Local leader-led challenge to make a small initial change (improvement)—learning stage;

Decade II: 1997–2007. Zero-to-One Project (Stage 1) designed and implemented for motivated village communities—first proposal and implementation;

Decade III: 2007–2017. Zero-to-One Upscaled Project (Stage 2) designed and implemented for motivated valley-wide village communities—second proposal and implementation;

Decade IV: 2018 onwards. Project yet to be self-developed by community people.

Specifically, what is the Zero-to-One community vitalization movement? This movement has two aspects: (1) design of a SMART governance scheme for communities motivated to pursue transformation and willing to experiment with ways to change; and (2) within this framework, each participating community accepts the challenge to become engaged in their own vitalization project by proposing and implementing small-scale, self-governance actions. SMART is an acronym: "S" represents small-sized and survivability-minded, "M" modest-scale and multiple-stakeholder involved, "A" anticipatory and adaptive, "R" risk-concerned and responsive, and "T" transformative.[2]

The name and concept of the Zero-to-One Movement was originally proposed by the author and then adopted by the Chizu Creative Project Team (CCPT), a resident-led voluntary group. Zero-to-One Movement indicates that the movement is anchored to and starts from an extremely small, easily understandable and manageable step that leads gradually to a change-making practice that can recover and revitalize a community's identity (thus getting to "one"). This is an essential process of *machizukuri*, particularly in rural communities that have been stressed by loss of population and a decrease in coping capacity. In 1996, the scheme was proposed to the then Mayor of the town by the CCPT and finally approved by the local Council.

Specifically, the Zero-to-One self-governance project requested participating communities to adopt and comply with the following three principles to win back the community's self-governance capacity:

(a) Make intracommunity communications more open and promote communication with outside communities;
(b) Reestablish local pride and autonomy; and
(c) Undertake community resource development and management.

Participating communities are qualified to join the project only if they have successfully set up their own project promotion associations (Zero-to-One Community PPA) within their communities. The level of this qualification condition was set higher for the Zero-to-One Upscaled Project (Stage 2), because all communities located in the same valley had to agree to form a joint project promotion association. At the initiation of Stage 2, the participating communities in the same valley (called district) had already set up their self-governance body, Zero-to-One District Community PPA. Importantly, with such PPAs, people now have an institutional platform to become engaged in the small-change (transformation) project (for this Zero-to-One Movement approach see also Sugiman, 2006).

5.3.2 Major Outcomes and Findings

Administrators, academics, other experts, and local residents attended annual evaluation meetings. So far, major visible outcomes included the reestablishment of a self-governance body at the valley level and the successful reuse of abandoned elementary schools for the purpose of a local-cuisine eco-restaurant, office spaces for small industries, and a local pension business startup for community residents, among others. New lifestyles have also been introduced in the town, such as new residents starting a *Waldkindergarten* (German for kindergarten in the forest) in Chizu

---

[2] SMART governance is differentiated from "smart" in the sense of smart technology or smart grid design.

Hang up buy Gov't's Crane   Self-floating by Zero-to-One Balloon

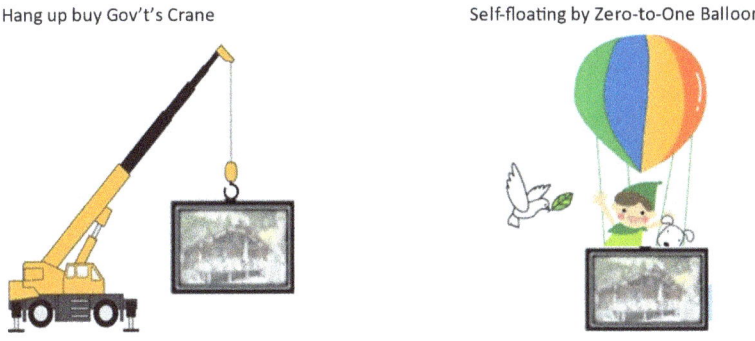

Fig. 5.6 Conventional top-down rescuing approach by the government like a lifting crane (left)
versus the bottom-up approach by the Zero-to-One scheme like a rising balloon (right)

Town, running an ecofriendly bakery and a brewery using fermentation with wild
yeast, and a restaurant in a renovated former kindergarten.

Figure 5.6 compares the conventional top-down approach (left) and the Zero-
to-One Movement project bottom-up approach (right). The conventional top-down
approach repeatedly rescues local communities suffering from rural-decline stressors
(due to loss, an aging population, and other macrosocial phenomena), diminishing
the coping capacity of communities, worsened by the loss of pride and hope in living
in their communities. In contrast, the Zero-to-One *machizukuri* scheme is designed
to provide the special leverage needed for a small transformation to be adaptively
promoted by the respective communities themselves.

Another significant outcome of the three decade-long *machizukuri* practices in
Chizu is the scientific observation and evidence-based field research conducted
by researchers and students. These research-education combined activities were
smoothly performed on the basis of the common platform that has been gradually
built up in the process of social implementation.

Figure 5.7 illustrates an example from Hayase village community which is a part
of Chizu town where the Zero-to-One common platform was effectively created by
using the Yonmenkaigi System Method (YSM), a unique workshop method originally
developed in Chizu by Okada and others (Okada, Fang et al., 2013a, 2013b; Okada,
Na et al., 2013a, 2013b). In this case, the stakeholders included the local leader, local
elderly residents (traditionally mostly males but in this project also females), local
non-elderly residents including females and children, and local government officials
(as advisers), together with researchers (as observers and advisers) and students
(assistants). In other Zero-to-One project communities, similar common platforms
have been strategically activated by use of the YSM, which has been commonly
implemented in multiple project areas as a testable model.

**The Yonmenkaigi system (YSM), originally designed and used for collaborative
action development for small groups in community-citizen vitalization initiatives
(Machizukuri) in a mountainous area of Chizu Town, Tottori, Japan**

Fig. 5.7 Chizu Town's challenges to build a common platform with the use of the Yonmenkaigi
System Method (YSM) – Hayase village community, Zero-to-One Stage 1[3]

The upper-left photo shows the scene of community people participating in a
workshop by use of the Yonmenkaigi System Method (YSM). The YSM chart-based
community plan developed by the people is shown in the upper-right photo. After
building up together a small community rest area, the community people posed for
the lower-left photo. The lower-middle photo shows the community vision painted
and developed together by the village people of Hayase, Chizu Town. In the lower-
right photo People were working together to pack up their local product according
to their community plan.

5.3.3 People's Coping Capacity Challenged by the Western Japan July 2018 Heavy Rainfall

During the latest Western Japan Heavy Rainfall Disaster (*The Asahi-Shinbun*, 8 July
2018), Chizu Town also suffered an extraordinarily heavy rainfall disaster over three
days, with more than 500 mm rainfall. This was an extreme torrential rainfall on an

---

[3] Photos: Courtesy of Mr. Shoichiro Nagaishi (upper left, taken in December 1996), Mr. Jon-il Na
(upper right, November 2006), Mr. Shoichiro Nagaishi (lower left, June 1997), Mr. Jon-il Na (lower
middle, November 2006), and Mr. Shoichiro Nagaishi (lower right, November 2006).

unprecedented scale, even exceeding the previously worst-case scenario of the 1961 Muroto Typhoon Disaster.

A quick field survey by the author showed that there was no loss of life in any of the villages that have been adaptively transforming their communities to more self-governable ones under different natural and social stressors. People seem to have evacuated in a timely fashion or at least decided to be on alert to evacuate together. In some communities, people made the decision to go to their nearby disaster evacuation shelters instead of running to a more distant public evacuation shelter designated by the town. The Mayor of Chizu took a strong initiative by loudly shouting into a microphone and directing people to be aware that the situation was just unbelievably dangerous and people could lose their lives if they did not evacuate voluntarily. The Box 2 shows some quotes (or paraphrased comments) from the local people interviewed by the author.

Box 2: Quates from the local people

- **Regarding evacuations**

  "As we found it wouldn't be safe to run to the public shelter, we decided to evacuate together to our nearest community center. It was a good choice because the designated public shelter was rather distant and we were not sure what would happen on our way."

  "In our community the situation was worse. Someone reported to us the road to the public shelter was already flooded. This news was immediately shared by everyone in the community and thus we made a choice to go together to our community center for evacuation.""We knew some elderly people in our neighborhood live just by themselves, so we ran to them and helped them to move to the shelter well in advance. Then some younger people stayed in our home, just ready to evacuate." "Some of us took good initiative to judge where to go and guide people."

- **Regarding the Chizu Town Mayor's "Evacuate immediately message"**

  "Every one of us immediately recognized his voice and the tone was very special indicating the situation is just extraordinary." "Maybe that was the strongest sense of crisis the Mayor conveyed to us timely and effectively!" "People say he is a very good communicator."

- **Regarding the Zero-to-One Movement**

  "Without the Zero-to-One movement, we could have coped with this abnormal situation much less effectively and timely."

  "The Zero-to-One Project Promotion Association (PPA) certainly gives us a better way to work together for a small change, step by step, and issue by issue. Likewise, to prepare for and fight with disaster PPA works a lot."

  "Our past accumulated experiences through these movements must have helped us increase our coping capacity as a whole."

On the whole, people's coping capacity has been found to work relatively well in the event of a large-scale disaster. A community's coping capacity can be adaptively developed through a variety of ways of handling day-to-day small issues, as well as through learning to fight in the face of natural hazard-induced disasters and social decline issues. As agreed by the village people interviewed by the author and as reflected by the above quotes from the people who experienced the latest heavy rainfall, without their Zero-to-One Movement achievements, the result could have been much worse.

5.4 Adaptive Process and Adaptive Design Framework for SMART Governance

5.4.1 Modeling the Zero-to-One Movement Process as an Adaptive Process for SMART Governance

As mentioned in the above, Okada (2017) proposed to model the dynamic process of the Zero-to-One Movement as an adaptive process for SMART governance. This notion of an adaptive process for SMART governance was developed by the author based on the methodological approach called (active) adaptive management (Folke et al., 2012; Holling, 1978).

The five features of SMART governance are proposed by the author as essential characteristics of the Zero-to-One Movement. There could be variants of these essential features.For instance, one might include self-solidification to the sublist of "S"— "Self-solidification" means that community people take the initiative to concentrate on specific plans, themes, or approach in their adaptive process. Or one could add reflective to the sublist of "R." As it stands now, it is largely arguable how rigorously one could define each of the essential features and their characteristics. This is particularly the case for transformative given that the theme itself entails edge-cutting research frontier issues that are challenged by an innovative approach (see IASS 2018, and Folke et al., 2012).

From the viewpoint of conducting implementation-oriented field research, it is also important to characterize the Zero-to-One Movement as a strategic approach to set up a common participatory platform that remains over a long period of time. This scheme is called "Case Station-Field Campus" (see Sect. 5).

5.4.2 A More Generalized Adaptive Design Framework: SMART Governance to Survive Persistent Disruptive Stressors

Based on the field-based experience and knowledge development in Chizu, Japan, a further challenge is to develop a more generalized adaptive design framework for implementing SMART community governance under combined natural and social risks. This framework is intended to be implemented in different communities of different countries, in order to design and actualize transformation for more sustainable communities.

At stake are innumerable small communities that will have to survive persistent disruptive risks (PDRs), including the external dynamics of demographic trends, mega disasters, and climate change.[4] As the 2018 Western Japan Heavy Rainfall Disaster shows, natural hazard-induced disasters tend to occur more frequently in a combined manner (such as typhoon, extreme rainfall, landslide, and so on) apparently under the influence of climate change. Local impacts of globalization and an interconnected world also act as persistent disruptive stressors (PDSs).[5] The challenge is how communities are to survive. Under PDS, they need a creative process and a common communication ground, whereby they adaptively learn together with respective stakeholders, transforming their lifestyles and social systems step by step, and eventually establishing SMART governance at the local community level.

As visually conceptualized in Fig. 5.8, the community at stake has been suffering and to suffer from a series of PDSs over a long period of time, say, decades-long time span. Okada has proposed to methodologically frame SMART governance within the decades-long perspective on the "community under PDSs." As already mentioned in

---

[4] The Intergovernmental Panel on Climate Change (IPCC, 2012, p. 1) states: "Extreme weather and climate events, interacting with exposed and vulnerable human and natural systems, can lead to disasters. This Special Report explores the challenge of understanding and managing the risks of climate extremes to advance climate change adaptation. Weather- and climate-related disasters have social as well as physical dimensions. As a result, changes in the frequency and severity of the physical events affect disaster risk, but so do the spatially diverse and temporally dynamic patterns of exposure and vulnerability."

[5] In the author's research work at the Institute for Advanced Sustainability Studies (IASS), the author has proposed to introduce the notion of persistent disruptive stressors (PDS) to mean basically an equivalent of persistent disruptive risks (PDR) as used in this article. From the viewpoint of a particular living community in focus, and from its residents' concerns, as compared to PDR, PDS might work better for them as the latter sounds more straightforward for people to imagine mega-scale dynamics with a huge potential to force them to survive and change. To be more consistent with risk research terminology, however, more appropriate would be PDR that addresses persistently revisiting the external risks to the community at stake. In conducting transdisciplinary research for some governance purpose, different terms and usages in different disciplines are often different in meaning and thus confusing. Experts' language and people's language are also not always mutually comprehensible. In addition, when we take such a challenge as mentioned here, we sometimes need new expressions or concepts, which add to the communication difficulties. The PDR versus PDS arguments thus lead all concerned to face such challenges in a particular context. When we wish to take a further step towards the social implementation of our research, this is exactly what challenges us.

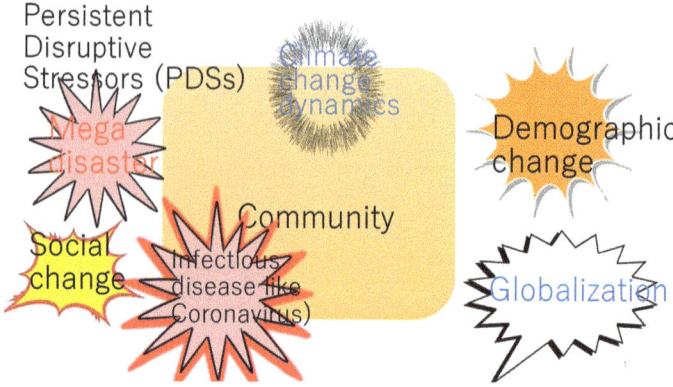

Fig. 5.8 Community under persistent disruptive stressors (PDSs)

Introduction, the term "disruptive" is introduced to indicate some possible (positive or negative) dynamic impact that results in a (small or large) change in quality or structure of the society or organization under that sort of external risk. Therefore "disruptive" is meant to be different from "destructive." Importantly, "disruptive" means not just deconstructive but it implies opening for innovation (Christensen, 2015).

Figure 5.9 shows that another framework to combine with SMART governance under PDSs is to extend the conceptual processes of "Build Back Better" as presented by UNSDR's Sendai Framework for Action which was formally adopted a principle after the Great Tohoku (Eastern Japan) Earthquake on March 11, 2011. During the negotiation period for the Sendai Framework, the concept of "**Build Back Better**" was proposed by the Japanese delegation as a holistic concept which states: "The principle of '**Build Back Better**' is generally understood to use the **disaster** as a trigger to create more resilient nations and societies than **before**. Note that the term resilient is used as a trigger to create a better society in the holistic sense which goes beyond disaster risk reduction.

This means that the concept of Build Back Better ("BBB") is rather a dynamic one than static, namely just to remain safe, secured and unchanged. Okada has proposed to extend this concept such that the region, city or community at stake should pay more attention to "before the disaster" and also "broaden the scope of agenda to include a whole set of concerns and interests to residents and other stakeholders." With this perspective in mind, Okada called the extended framework "Build Back Better even Before Disaster (BBBB)." If sustainable future is more strategically expressed when discussing BBBB, the framework may better be stated as "Build Back Better even Before Disaster towards Sustainable Future" or "Build Back Better even Before Disaster towards Better Being" (Note that "well-being" is another way to define sustainable states for the community at stake).

As illustrated by Fig. 5.9, Okada has developed a visionary model of the ups-and-downs processes of the community's wholistic state of sustainability "or its proxy

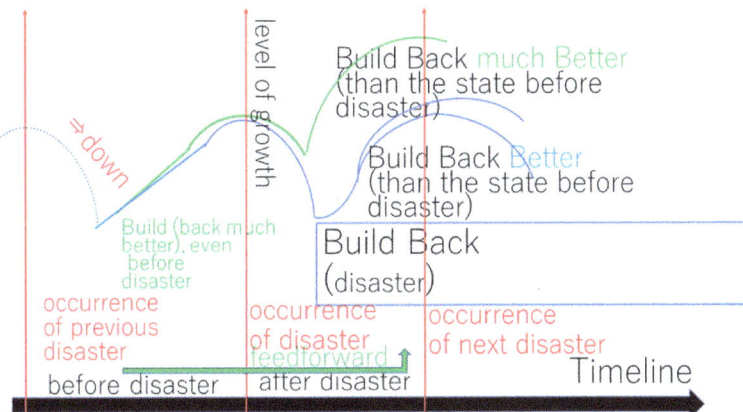

Fig. 5.9 Comparing the processes of build back, build back better, and build back better even before disaster (BBBB)

performance indicators as observed over decades-long time spans. This conceptual model may also serve for the purpose of illustrating and designing BBBB in relation to the SMART governance under PDSs.

5.5   Communicative Spaces for Dynamic Resilience

5.5.1   *For the Above SMART and BBB or BBBB Process Models to Work Effectively*

It is to note: if a "common ground" (or "communicative space") is more strategically set up, nurtured and long-maintained, the more successful and viable would become the dynamic processes of series of adaptation and transformations to take place. Okada and Goble (2021) proposed the following list of attributes to define and characterize "communicative spaces."

(1)   An auxiliary space annexed to its home-ground (i.e., the community at stake); space is either physical and/or virtual.
(2)   To be initiated by concerned actors (agents) first from its home-ground and joined also by outside actors
(3)   With a special purpose to be achieved (fulfilled)
(4)   Participating actors repeatedly meet and communicate in order to perform SMART Governance.
(5)   To be synergetically and communicatively developed by participating actors Just reversely, participating actors tend to be energized by the synergy of the space.

(6) This auxiliary space which is annexed to the original home-ground can possibly span over and become networked to counterpart auxiliary spaces outside of the home-ground and annexed to other home-grounds.

This means this common space which is annexed to the community (as a system) is conceived as an embedded sub-system to the system. If the purpose of the space is fully achieved, it may naturally dissolve itself.

There is a good piece of evidence from the case of Chizu to illustrate the points of dynamic resilience as shown in Fig. 5.10. This figure schematically sketches possible synergetic processes of Space x Process Dynamics, all of which contribute to the growth of dynamic resilience for the community at stake. Let us revisit **5**, where we studied how people in Chizu coped with the Western Japan July 2018 Heavy Rainfall. Thanks to a series of community's efforts to enhance their coping capacity to survive different PDSs towards sustainable future, they had already been prepared for such a surprise heavy rainfall disaster. As a result, village people proved to over-come this type of unprecedented scale of natural disaster and successfully increased their resilience against it. These entire processes accompanied by the successful evacuation are the real evidence of transformative dynamics which indicated the dynamic resilience exercised by the community people. They had nurtured step by step their communicative spaces for their active engaged in the Zero-to-One Move-ment commitments. At the same time, their coping capacity had been adaptively enhanced and dynamic resilience actualized by making use of the growing dynamics of the communicative spaces. Therefore, we may well interpret the whole progress as synergetic dynamics of both space and process, where communicative interactions are also working among people meeting together there.

Finally, we will just offer two examples of communicative spaces which have been implemented as outputs of SMART Governance in Chizu.

Fig. 5.10 Schematic sketch of synergetic processes toward dynamic resilience

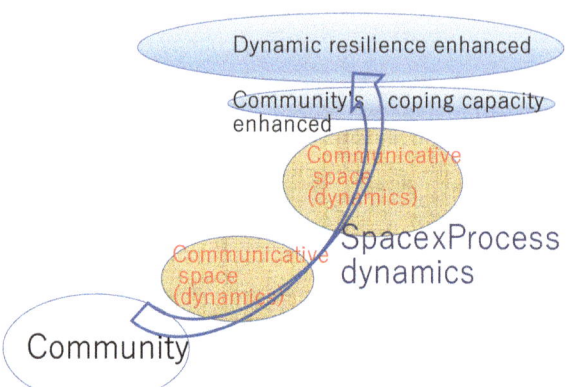

5.5.2 Communicative Spaces Strategically Built into the Zero-to-One Projects

As has been demonstrated through the Zero-to-One Projects introduced in Chizu for two decades, we may interpretate its unique, essential scheme was designed to provide a communicative space for the community and get them committed to developing and implementing small changes over a decade long time span.

We should remark that before they applied for this town- grant project, their communities simply kept missing such self-governance function, although they had traditional, routine type of community management community association. This explains well how it is instrumental for rural communities under PDSs to restore a communicative space for people to meet and come up with self-motivated small changes to survive and restore towards their own-visions set by themselves. We note that in this case of the Zero-to-one project, stakeholder representation (combined with willingness to collaborate) is what is needed. Also, stability and established rules, and legitimate processes are prime attributes.

5.5.3 Collaborative Research and Education Schemes based on the Case Station-Field Campus (CASiFiCA) Scheme

A unique and creative resident-led local initiative with strategic objectives and joined by researchers and students is called the "Case Station-Field Campus (CASiFiCA)" (Okada & Tatano, 2008). This scheme was first empirically evolved through the above-stated Zero-to-One projects in Chizu. Based on this protocol, multiple aspects of knowledge development are systematically pursued. The primary outcome is what we can call an "adaptive process design for sustainable communities."

Acknowledging that diverse efforts have been made for disaster reduction, particularly in disaster-prone areas (countries), many professionals have been energetically and devotedly engaged in fieldwork to reduce disaster risks. They also recognize that more community-based, stakeholder-involved approaches are needed. A crucial question is why we cannot conduct fieldwork more creatively. One promising arrangement to enable us to establish a common communication platform is the CASiFiCA scheme, originally proposed by Okada and Tatano (2008). The CASiFiCA scheme is characterized by a set of local research case study stations and field campuses (Fig. 5.11). Research case study stations (case stations, for short) are sites at which continuous case studies are conducted strategically by researchers and students stationed in the real field(s). In such real fields local stakeholders are engaged in cases to make a modest change towards betterment of their communities, and researchers and students conduct evidence-based studies, as if the fields were like an academic campus where multilateral knowledge of social implementation is scientifically documented and examined. For this reason, it is called a "field campus."

A Case Station and Field Campus (CASiFiCA) is set up as a unit of common platform for field-based collaborative research and education. To promote comparative case studies for different fields with testable models and practices, these units are networked together, across communities in the same region or country or further extended to include global communities. They are called the "CASiFiCA Network."

CASiFiCA has proven successful in promoting IDRiM (Integrated Disaster Risk Management Society) research and education activities. For example, Kumamoto University has successfully taken the initiative to develop a unique multiple-university collaboration disaster education program based on this CASiFiCA scheme (Okada, 2018b). The initiative also resulted in a new research-education human network, as well as a disaster relief and support communication platform by involving local administrators, the local Red Cross, the local meteorological observatory, local media, and other nongovernmental organizations. The April 2016 Kumamoto Earthquake actually challenged this platform, which was at an early stage of development. Although it was not yet operating at full scale, there was visible evidence of relief support initiatives that were self-organized in a timely manner by students and researchers from the network. Kumamoto University did not lose time to set up a field campus-research laboratory as an outreach to the devastated town of Mashiki, south of Kumamoto City.

The University's disaster reduction-related institute has expanded at the entire university level, now covering water, environment, and disaster reduction in a more integrated manner. The Kumamoto-based Case Station-Field Campus activities are expected to enhance their efforts to connect with other regions, including the Chizu Case Station-Field Campus activities.

In this relation, it may deserve attention that The Institute for Advanced Sustainability Studies (IASS) has been operating the KLASICA project, headed by Chabay (2017), as an international research alliance. Although KLASICA and CASiFiCA have been developed quite independently from each other, both are aimed at understanding and supporting the development of behavior and attitude change through community processes. A distinct value of KLASICA is its specific concerns with

Fig. 5.11 The Case Station-Field Campus (unit) scheme to support collaborative research

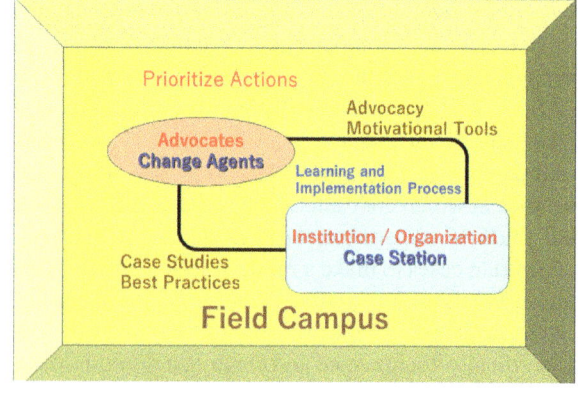

implementation processes and its methodological focus on narratives and affective expressions as an approach to understanding behavior change in communities.

With regard to a reflection of special experience and knowledge and more common experience and knowledge, there remains much room for both approaches to be cross-examined and, if appropriate, synergetic coordination of the two approaches should be considered in the near future. The proposed model of SMART governance to survive under PDR (or PDS) is expected to provide an effective framework for the Case Station-Field Campus (CASiFiCA) scheme. By networking CASiFiCA for different case areas, strategic comparative studies should be conducted in a transdisciplinary manner.

5.6 Conclusion

This Chapter discussed community resilience from the viewpoint of community's coping capacity challenged for surviving and regrowing towards sustainable future. Such a community challenge was called "rural decline" or more broadly "community decline" to refer also to urban areas. An integrative perspective has been introduced to characterize (rural) community decline. Rural decline is caused and accelerated by different natural and social stressors such as loss of population and aging, social and economic changes as well as natural disasters and climate changes. This was claimed to be a core challenge in addressing the rural decline occurring cross Japan.

The uniqueness and value of this study rested on the following two approaches. One was to provide conceptual and methodological frameworks to strategically build up dynamic resilience for the community at stake. The second was to make full use of field-based evidence accumulated through the author's three decade-long studies conducted in Chizu, Tottori, Japan. The case history which was built on the two approaches demonstrated that people have increased adaptively their coping capacity through the long years' participatory process, as an adaptive process for SMART community governance under persistent disruptive stressors. It can be argued that the two approaches are mutually supporting and consolidating each other.

Especially, to enable dynamic resilience, communicative spaces are critical. We have discussed possible synergetic processes of Space x Process Dynamics, all of which may contribute to the growth of dynamic resilience for the community at stake. Zero-to-One research project with Kumamoto University can be interpreted as a unique scheme already implemented to introduce and nurture communicative spaces. The Case Station-Field Campus scheme has also been proposed to serve as a common platform for studying adaptive design processes over a long period of time.

Though not discussed in this Chapter, more generalizations and comparative studies are needed. It is noted that the author and his research team have already performed such comparative studies (e.g., Iida City and its neighboring region of Tenryu River Basin). Just to mention a valuable outcome from the ongoing studies, both Chizu, Tottori and Iida, Nagano, Japan have experienced decades-long SMART governance practices which have been contributing to synthetic outcomes of sustainable developments such as poverty, good health, gender, quality education, climate change, forestry, water, etc. As a whole, theses progresses will demonstrate increasing performance of community's dynamic resilience.

The author has also started to work with communities in other countries with a view to examining the communality and limit of the proposed approached in this Chapter.

Acknowledgements The author wishes to thank Profs. Ortwin Renn, Ilan Chabay, Robert Goble and Mika Shimizu for their valuable comments and advice.

References

Cambridge English Dictionary. SYSTEMIC | meaning in the Cambridge English Dictionary.

Chabay, I. (2017). Overview of KLASICA: An IASS international research alliance. Presentation at the IASS-IDRiM Workshop on Risk Governance for Natural Disasters – Extending Integrated Disaster Risk Management to Sustainable Communities, 29 August 2017, Institute for Advanced Sustainability Studies (IASS), Potsdam, Germany. https://www.iass-potsdam.de/en/research/kno wledge-learning-and-societal-change-alliance-klasica. Accessed 4 September 2018.

Christensen, C. M., Raynor, M. E., & McDonald, R. (2015, December). What is disruptive innovation? *Harvard Business Review*. https://hbr.org/2015/12/what-is-disruptive-innovation. Accessed 17 November 2018.

Folke, C., Carpenter, S., Elmqvist, T., Gunderson, L., Holling, C.S., Walker, B., Bengtsson, J., Berkes, F., et al. (2012). Resilience and sustainable development: Building adaptive capacity in a world of transformations. Environmental Advisory Council, Ministry of the Environment. http:// era-mx.org/biblio/resilience-sd.pdf. Accessed 17 November 2018.

Holling, C. S. (Ed.). (1978). *Adaptive environmental assessment and management*. Wiley.

IASS (Institute for Advanced Sustainability Studies). (2018). Website. https://www.cigionline.org/ partner/institute-advanced-sustainability-studies-iass. Accessed 22 November 2018.

IPCC (Intergovernmental Panel on Climate Change). (2012). Managing the risks of extreme events and disasters to advance climate change adaptation – Special report of IPCC. https://www.ipcc. ch/pdf/special-reports/srex/SREX_Full_Report.pdf. Accessed 17 November 2018.

Kawahara, T., & Sugiman, T. (2003). Revitalization of a rural depopulated community by introducing participatory democracy system into each smallest unit of community. *The Japanese Journal of Experimental Social Psychology, 42*(2), 101–119.

Okada, N., & Sugiman, T. (1996). A research perspective of rural declining towards the development of community planning research. *Journal of Japan Society of Civil Engineers 562*(IV-35), 15–25 (in Japanese).

Okada, N. (2006a). Perspective on integrated disaster risk management. In Y. Hagihara, N. Okada, & H. Tatano (Eds.), *Introduction to integrated disaster risk management* (pp. 9–54). Kyoto University Academic Press. (in Japanese).

Okada, N. (2006b). Urban diagnosis and integrated disaster risk management. *Journal of Natural Disaster Science, 26*(2), 49–54.

Okada, N., & Tatano, H. (2008). Case Station-Field Campus (CASiFiCA): Globally-networked, field-based research and education challenges for disaster reduction. Presentation at the International Disaster and Risk (IDRC) DAVOS Conference 2008, 25–29 August. https://idrc.info/fileadmin/user_upload/idrc/former_conferences/idrc2008/presentations/Program_forWebsite_V7.pdf#search='Davos+conference+Case+StationField+Campus+%28CASiFiCA%29+%3A+Globallynetworked%2C+Fieldbased+Research+and+Education+Challenges+for+Disaster+Reduction. Accessed 20 July 2018.

Okada, N. (2010). City and region viewed as Vitae system for integrated disaster risk management. Annuals of Disaster Prevention Research Institute, Kyoto University, 49 B: 131–136.

Okada, N., Na, J., Fang, L., & Teratani, A. (2013a). The Yonmenkaigi System Method: An implementation-oriented group decision support approach. *Group Decision and Negotiation, 22*(1), 53–67.

Okada, N., Fang, L., & Kilgour, M. (2013b). Community-based decision making in Japan. *Group Decision and Negotiation, 22*(1), 45–52.

Okada, N. (2016). Who and how triggers and drives resident-led participatory planning and management for better living environment - Japan's "Machizukuri" challenges. In B. A. Misra, B. N. Singh, H. B. Singh, and Kusuma Lata (Ed.), *Indian urbanisation and sustainable development: Spatial perspectives*, pp. 101–115. Kanishka Publishers.

Okada, N. (2017, January 16). Beyond disaster risk reduction: Smart governance to survive Persistent Disruptive Stressors (PDS). Final Seminar at the Institute for Advanced Sustainability Studies (IASS).

Okada, N. (2018a). Adaptive process for SMART community governance under persistent disruptive risk. *International Journal Disaster Risk Science, 9*, 454–463.

Okada, N. (2018b, April 17–18). Research networking by Field-Campus and Case-Stations (CASiFiCA) for "Build Back Better." Presentation at the Asian Science and Technology Conference for Disaster Risk Reduction.

Okada. (2021, November 26). Another challenge: Systemic thinking and design for community-based/humans-focus approach, Invited Speech at DPRI SOGO-Bosai Seminar 50th Session.

Okada and Goble. (2021, March). Maintaining vigilance is critical and challenging for disaster risk management: It offers lessons in implementation science Part 1 and Part 2; The 5th International Symposium on Natural and Technological Accident Risk Reduction at Large Industrial Parks.

Sugiman, T. (2006). Theory in the context of collaborative inquiry. *Theory & Psychology, 16*(3), 311–325.

The Asahi Shinbun. (2018). Death toll hits 81 as western Japan braces for more heavy rain. 8 July 2018. http://www.asahi.com/ajw/articles/AJ201807080023.html. Accessed 15 August 2018.

United Nations. https://www.undp.org/sustainable-development-goals. Accessed 20 July 2018.

Norio Okada is a Professor Emeritus of Kyoto University and currently an adviser to Institute for Disaster Area Restoration, Regrowth and Governance, Kwansei Gakuin University, Japan. He graduated in 1970, received M.Eng. in 1972 and Dr, Eng. in 1977 all from Faculty of Engineering, Kyoto University. His academic background is civil and environmental engineering, mainly mathematical modeling and systems approach to infrastructure planning and management, disaster risk management, and community-based participatory approach. He took the initiative to launch the international society of Integrated Disaster Risk Management (IDRiM). He has edited and published many books such as *Water and Disasters* (co-edited by Gopalakrishnan and Okada, Routhledge, 2007).

Chapter 6
Collaborative Dialogue-Based Approach for Environmental and Disaster Resilience and Governance in Japan

Hidenori Nakamura

Abstract In this chapter, I use the two field explorative case studies in post-disaster Japan, i.e., local stakeholder dialogue on volcanic disaster management, and citizen dialogue with experts on radioactive waste, to propose relevant methods to develop a culture of dialogue that nurtures resilience, in the context of two complexities: complexity of scientific knowledge, and complexity of people's lived experiences and value systems. The visions and methods described here are relevant to a resilience approach in that they focus on boundary and creating environment by linking scattered resources, looking at details and the whole. The methods proposed are intended to lower the walls, connecting the dots unconnected so far, to change the perceptions and lived experiences of boundary of partnership, in the context of local policy process and collaboration. Drawing the local case study, I argue that researchers could play significant roles in science-based policy process that engages local stakeholders and general public, changing the beliefs of boundary between experts, practitioners, and citizens. The case study also suggests that explicit treatment of pluralistic positions and views among researchers and stakeholders would reduce the barrier of quality learning and dialogue in science-based policy process.

6.1 Introduction: Two Series of Dialogical Meetings in Social Networks in Japan as a Demonstration of the Resilience Approach

To meet the United Nations Sustainable Development Goals (SDGs) by creating sustainable and resilient societies, an effective way to manage scientific knowledge in the context of the complexity and uncertainty in contemporary society is required. Here, effectiveness refers to the degree to which the means adopted actually contribute to realizing the intended ends. Drawing on the case of Japan's environmental and disaster policy research, this chapter

H. Nakamura (✉)
Department of Environmental and Civil Engineering, Toyama Prefectural University, Imizu, Japan
e-mail: hdnakamu@pu-toyama.ac.jp

argues that researchers could play significant roles in science-based policy processes that engage local stakeholders and the general public. Strategic actions by researchers adopting a resilience-based approach can break down the walls between experts, practitioners, and citizens. Such an approach addresses the interlinkages between stakeholders, as well as issues/risks, natural/social/human systems, resources, and generations (see Chapter 1). The approach is expected to develop an environment in which stepwise perceptional change can occur. The scientists' contribution to open dialogue with stakeholders in a policy process can help overcome positionality and scientific uncertainty constraints. Positionality refers to organizational straitjackets that regulate organizational members' behavior so that they only represent organizational interests (cf. veil of ignorance, Rawls, 1999). This dialogical environment could change how SDGs are perceived, not scattered and independent, but integrated and interconnected, and help materialize partnerships toward the goals.

The following text provides two field explorative case studies, that is, local stakeholder dialogue on volcanic disaster management, and random-sampling based citizen dialogue with experts on radioactive waste, to propose suitable methods to develop a culture of dialogue that nurtures resilience in the context of two complexities: (1) complexity of scientific knowledge, and (2) complexity (diversity/inconsistency/difficulty of sharing) of people's lived experiences and value systems.

The ideas, visions, and methods developed in the studies are relevant to a resilience approach in that it focuses on boundaries, missing links, or creating an environment by looking at details and the whole (*tree and forest* as a metaphor) in a continuum (See Chapter 1 and 2). By experimenting with the possibility of alternative beliefs and perceptions in a psychologically safe environment, we are more likely to challenge conventional views and believe realities in a less coercive manner. When a change is needed to achieve SDGs, the degree of coercion in achieving the goals is also an issue. The subjective beliefs and perceptions, or unconscious mental models, shall be explicitly and carefully identified to enable dialogue and collaboration across borders and walls of positions and interests. Lowering walls requires the temporal cessation of reproducing familiar mental models. Giving up applying one's own view for a moment may imply the perceived removal of armor, reducing the sense of safety. Therefore, an environment in which people can be vulnerable enough to be dialogical or open to other views is necessary. The metaphor for open dialogue is to cross the border and open the way we are, in a non-coercive way.

Two post-disaster case studies in Japan are presented in this chapter. One is the three-year capacity development of local volcano disaster management in three prefectures in central Japan (a prefecture is equivalent to a state or province in other countries) after the Mt. Ontake eruption in 2014 (Nakamura et al., 2019). The other is a five-year long random-sampling citizen dialogue on the geological disposal of radioactive waste, held in Aichi and Shizuoka prefectures in central Japan and Fukushima prefecture after the Fukushima Dai-ichi nuclear power plant accident in 2011 (referred to "the 2011 Fukushima accident" hereinafter) (Nakamura et al.,

2021). In the former case, researchers functioned as producers, coordinators, and facilitators for capacity development, collaborating with local government officials. In the latter case, researchers played the role of scientific interpreters with differing views, reflecting dialogue among citizens. These studies suggest that explicit treatment of pluralistic positions and views among stakeholders and researchers is a key for quality learning and deliberation in a policy process, which changes the perceived barrier of partnerships and interlinkages, that is, who are the fellows (individuals with whom one could collaborate), and what are relevant/irrelevant issues and values to be addressed. Here, the fellow is neither a friend nor enemy. These perceptions of fellows as well as friends/enemies create a barrier of collaboration (and competition and domination, more specifically, the relationship between the dominator and the dominated). Moreover, the judgment of relevance/irrelevance in subjective belief construction also induces another barrier of collaboration.

The findings of citizen dialogue are also about the explicit treatment of inclusiveness in the participatory policy process. The perception of inclusiveness allows us to go beyond the wall of the perceived roles of experts and citizens: (1) citizens are not necessarily the last to be involved and valued in policy processes, and (2) experts are not necessarily the first to be involved and valued in policy processes. Inclusiveness is part of SDG-relevant vision, which shall be realized, not only as a goal or state to be measured but also as a process or way of manifesting the vision.

As a concluding perspective and a focal point for further research and practice, I discuss three points that need to be addressed in collaborative partnership development: (a) border and membership (or fellowship) recognition of collaboration, (b) positionality (relevant to a sense of community belonging to perceived / subjective subsistence), and (c) disclosure and sharing of weaknesses and fear. This is about the less coercive change of perceived border, or mental model, about who we are and what we can and should do, using open dialogue in inclusive policy processes.

6.2 Volcano Disaster Case

6.2.1 Clinical Environmental Studies: A Trans-Disciplinary Approach

To conduct trans-disciplinary research on volcano disaster management in Japan after the Mt. Ontake eruption in 2014 (the worst volcano disaster in Japan since 1945, causing 58 casualties and five missing climbers), researchers, including the author of this chapter, applied the perspective and method of clinical environmental studies. This is a methodology of trans-disciplinary actions for better sustainability research and education, recently developed by Japanese academics in partnership with graduate students and local stakeholders. Here trans-disciplinary actions refer to collaboration between researchers and practitioners/citizens (Kato et al., 2014). The fundamental concepts of clinical environmental studies include (1) diagnosis and (2) treatment, where the researchers would progress from the diagnosis of the sustainability problems to the treatment stage, with the attitude and capacity of (3) interdisciplinary research from various areas, and (4) trans-disciplinary actions with non-research partners (Nakamura et al., 2019). In the diagnosis stage, the researchers develop a (5) problem map, which describes the hypothetical elements and their causality/relationship with the problem and the whole systemic picture, in collaboration with stakeholders. This approach stresses the spiral execution of diagnosis and treatment as a collaborative and social learning process, naming (6) continuous working hypothesis development. Here both diagnosis and treatment are recognized as provisional, and are subject to revision based on the joint and shared observation and experiment.

As an application of the clinical environmental studies approach, a three-year research project on volcano disaster management and resilience capacity development was designed and implemented as the first cycle of diagnosis and treatment.

6.2.1.1 Executing Network of Practitioners and Researchers

The project was led and coordinated by a core team of researchers and practitioners from a university and Gifu prefecture in Japan. Nagano and Ishikawa prefectural governments were also responsible for disaster management of the three target volcanos, Mts. Ontake, Yakedake, and Hakusan. All three volcanoes are situated in central Japan and on the borders of prefectures. Moreover, each attract climbers; however, unlike the other two, Mt. Yakedake is close to residential areas and is a popular tourist destination. Nine municipal governments in areas near the target volcanoes were also engaged in study meetings and workshops. Researchers from two other universities who had been studying two of the three target volcanos proactively participated and contributed to the project. As such, a network of researchers and practitioners implemented the project. All the local governments listed as members of the three volcano disaster management committees established by law were invited to participate.

Nurturing ownership, or a sense of agency, was one of the main capacity development goals at the local level. Project ownership was intentionally shifted gradually from the researchers to the staff of prefectural governments and then to the municipal government officials. The number of members participating in detailed planning and execution of activities was increased stepwise (Nakamura et al., 2019).

6.2.1.2 Stepwise Development of *ba*

The three-year development of *ba* (in Japanese), where dialogue is held between practitioners and stakeholders, as well as between practitioners and experts, was designed as follows (Nakamura et al., 2019). Here, *ba* is defined as "formal, non-formal, and informal institutional and/or organizational settings where tacit knowing is externalized among participants, leading to the succession and modification of shared recognition and behavior." *Ba* is a physical and non-physical setting of a shared time and space for emerging relationships among individuals and groups to create knowledge, exemplified by workshops and dialogical meetings. With reference to resilience, *ba* development was intended to observe and develop interlinkages between stakeholders, find and address missing links, and lower the walls between positions.

Every year, a joint study meeting and workshop (or symposium for the last year) was held, in which local government officials in charge of disaster management and designated researchers for three volcanoes were invited. Moreover, study meetings and workshops were held annually for each volcano, extending participation to include members of the volcano disaster management committees for each volcano for the first and last year and key local stakeholders such as residents, businesses, and non-profit organizations (NPOs) for the second year. The main theme of the workshops was problem identification and mapping among participants during the first and second years. For the final year, ideas of action plans after completing the project were shared (Nakamura et al., 2019).

6.2.1.3 Workshop Allowing Individuals to Speak and Listen, Detached from Organizational/Positional Constraints

The workshops were specifically designed to complement the current functions of the official volcano disaster management committees. These committee meetings used to be held once a year for several hours and included representatives from several member organizations. The agenda and proposals were prepared by the secretariat of the committee, that is, the headquarters and field offices of prefectural governments. In these formal meetings, it was impossible for everyone to have an adequate exchange of opinions and views in a limited time with such a large number of participants. Moreover, there was a tendency for official decisions to be taken based on the draft prepared by the secretariat without major changes. Therefore, substantial content was not developed during the committee meetings but was mostly added during the preparatory work by the secretariat members.

To achieve substantial communication among stakeholders who were not merely representing the positions of the organizations to which they belonged, the workshops were designed to allow participants familiar with contexts of the organizations to which they belonged to exchange honest views and ideas, for the purpose of exchange as an end in itself. There were no explicit requirements for these workshops to reach a conclusion or agreement among the participants. Moreover, three to six small groups comprising six to eight participants from different organizations or geographical areas of residence were organized to develop opportunities for face-to-face interactions among participants. Middle rank to junior officials of the steering team for each volcano committee were asked to be the facilitators to promote dialogue among those participants who did not know each other well. Each workshop was organized and coordinated by a researcher who was responsible for setting the tone for overall conduct after prior consultation on the program with core team members. Time management on the day of the workshop was made flexible and could be adjusted to the participants' feelings to ensure that the participant satisfaction levels were maintained. The design of the dialogues described here was introduced to cope with the positionality of individuals, which acts as a straitjacket for governing participants' thinking and behavior.

6.2.2 Outcome of the Trans-Disciplinary Research Project

Applying the trans-disciplinary approach known as clinical environmental studies, the three-year project aimed to develop the capacity of disaster management committees for three target volcanoes with a comparatively lower frequency and magnitude of volcanic activity in Japan. The core members intended to reinforce a sense of agency and ownership of disaster management by municipal government officials in charge of planning, coordination, and execution of the committees' activities. More specifically, it was intended that their willingness to continue *ba* development to nurture interpersonal and inter-organizational relationships among local stakeholders should be increased by the end of the project. The remarks of local government officials in the final workshop for each volcano committee suggest an increased sense of problem ownership: "It would be important for us to continue developing face-to-face, mutually supportive relationship further (Mt. Ontake)"; "To cope with the changing membership of the volcano disaster management committee, it would be effective for us to have relationship maintenance meeting once a year, exchanging expert views. (Mt. Hakusan)"; and "We should have an opportunity of free discussion, at least once a year. (Mt. Yakedake)" (Nakamura et al., 2019). However, cultivating a sense of ownership among local residents, businesses, and NPOs remains an issue since this project first focused on raising a sense of ownership among local governmental officials, connecting across the geographical and mandate borders, within limited project duration. Nonetheless, the engagement of residents and stakeholders other than government officials and experts is also important in terms of sustainable local development.

6.2.3 Value Systems and Practitioner-Researcher Relationship

To understand the perceived and persistent images and subjective realities at the border of collaboration, that is, those with whom one can collaborate, and to lower the border wall, perceptional change regarding who we are and what we should [can]/should [can] not do, is fundamental. Fixed understanding of "what I can and should" hinders experimentation with new collaborative actions beyond the existing roles and functions of the current stakeholders. The case study tackled the following action-research perspective-based questions. In order to improve volcano disaster preparedness and coping capacity in Japan, particularly for volcanoes with a relatively low frequency and magnitude of volcanic activity, (1) what kinds of value systems need to be changed, and (2) what kinds of relationships need to be changed? Open dialogue in the case study brought out some factors that might hinder collaborative partnership/social network development at the local level.

Concerning the first question, a primary factor is a belief in a particular position among professional individuals working for specific organizations. According to the core local government official member of the project, this becomes a problem for local government officials when they engage in coordination, collaboration, or consultation beyond their own geography and mandate boundaries to address volcano disaster risks that transcend boundaries. Therefore, a series of study meetings and workshops to nurture open dialogue, organized and coordinated by researchers, are needed to change beliefs that certain positions are inevitably embedded in certain people.

In addition, it is necessary to change value systems in terms of the role of local government, as well as that of local residents and businesses. The local government can no longer be expected to act as commanders and controllers of residents and businesses but rather as facilitators and stakeholder coordinators. As shown in the partnership vision in SDG 17, local residents and businesses are partners with local governments. They should see themselves as owners rather than subordinates of the local government. As such, dialogical meetings in workshops can create a horizontal relationship between the local government and both residents and businesses.

Regarding the second question, a primary factor is the relationship between local governments and universities (or practitioner and researcher). As demonstrated by the case study, university researchers could take the role of producers, coordinators, and facilitators of *ba* development in collaboration with the local government. Although universities are institutionally responsible for education, research at universities is basically conducted by each researcher, not by universities as institutions. This makes it difficult for universities to collaborate continuously with local governments or other external partners. However, a network of individual researchers can develop and maintain a collaborative platform with local governments and empower the relevant actors for disaster management.

As shown in the above case of local government (officials) and university (researchers), and the relationship between these two entities (professionals), the change in the perceptional border of who we are, and what we should [can]/should [can] not do, can be facilitated through dialogical meetings and *ba* development, with clear intention and awareness of experiment and observation. Self-design and reflection by (core) experiment participants and members are essential to realizing learning with open-mindedness and help developing tolerance to uncertainty.

6.3   Radioactive Waste Case

6.3.1   Developing Methods of Citizen Dialogue on Environmental Policy in Japan

After the 2011 Fukushima accident, an explorative study of citizen dialogue was conducted to cultivate a culture of dialogue as part of enhanced participatory environmental governance in Japan (Nakamura et al., 2021). A culture of dialogue is defined as an attitude that allows an individual to communicate with other individuals, even if they have different views. As specific methods to encourage dialogues to address the issue of hesitation regarding confrontation and pluralistic reasoning, the study proposed: (1) politeness-based facilitation dialogue, (2) evidence-based and position-explicit presentation of experts with different views, and (3) experts reflecting in tandem with citizens engaged in dialogue. The study also examined the effects of these methods on attitudes toward dialogue and internal self-deliberation about policy options.

6.3.1.1   Politeness-Based Facilitation Dialogue

First, a politeness-based facilitation dialogue was developed through a series of workshops to contemplate and realize sustainable development in Aichi, Japan, including the EXPO in 2005, referring to the Japanese concept of *jinen* (spontaneity). *Jinen* is a state in which a decision is made naturally in support of the unconscious, the body, and the self. Politeness-based facilitation dialogue has the following features (1) not denying other participants' remarks, (2) no one controls the dialogue process, and (3) there is no need to reach an agreement or conclusions. This method has been used as a part of the methods for facilitating citizen dialogue in Japan. The weak culture of dialogue and the emotion/value system to hesitate confrontation was taken into consideration.

6.3.1.2   Experts with Differing Views

Second, the method of interpretation that focuses on communication between experts (science and technology) and citizens was developed and applied in the case study of citizen dialogue. Experts with different views and ideas participated in citizen dialogue with rules of (1) evidence-based presentation and explanation and (2) explicit disclosure of one's own position, such as the assumptions underlying the conclusions. Based on the sociology of science of the 2011 Fukushima accident, a Japanese sociologist argued that revealing a multiplicity of experts' presumptions and positions (e.g., interpretation of low-level radiation health impacts) is significant in societal communication and decision-making in addition to evidence-based policymaking (Matsumoto, 2012). It is important to deal explicitly with pluralistic views to make collective decisions where no unique or optimal solution exists. Allowing for multiplicity among experts' views is a prerequisite for a less coercive policy process, in line with the concept of *jinen* (spontaneity).

6.3.1.3 Reflections by Experts in Front of Citizens

Third, the "open dialogue," a psychiatric medical system and treatment originating in Finland, was examined to see if it could be applied to citizen dialogue on policy issues related to sustainability. Open dialogue has been developed as an integrated system of care provision and a way of treating mental health issues in the Western Lapland area of Finland since the 1980s (Putman & Martindale, 2022; Seikkula & Arnkil, 2006). Compared to conventional systems and therapies, the open dialogue approach embraces the values underlying a culture of dialogue to deal with confrontation and pluralistic reasoning in a less coercive manner. The approach has shown better medical and social outcomes of patients such as lower requirements for hospitalization and medication, as well as better employment prospects after treatment for patients who were difficult to work due to mental health problems. The approach developed open dialogue principles based on research, which include tolerance of uncertainty and dialogism as two core principle.

Tolerance to uncertainty encompasses increasing psychological resources to cope with difficulties in constructing social realities with others, through experience sharing (no manipulation of situations), new social network creation, and whole-body listening and talking (binding the utterance with the embodied experience through emotions) among others (Seikkula & Arnkil, 2006). Dialogism is being dialogical, responsive and humble in the face of the unknown, maintaining relationships, and avoiding repetition of preset monologues.

The open dialogue approach first caught the attention of Japanese practitioners and researchers around the mid-2010s. A Japanese psychiatrist, who introduced this approach to medical treatment in Japan, and a Japanese sociologist, who studied the recovery process after the 2011 Fukushima accident and shared the results outside Fukushima and who also had the experience of facilitating dialogue on nuclear and radioactivity issues, suggested this approach and its effectiveness for dialogue between pro-nuclear and anti-nuclear people in Japan after the 2011 Fukushima accident (Saito & Kainuma, 2016).

According to the suggestion of the above psychiatrist, the method of "reflecting" in the open dialogue approach was proposed in the case study of citizen dialogue (Saito & Kainuma, 2016). "Reflecting" is an opportunity for a care service user and his or her network members to observe medical experts' mutual discussions in front of them in the context of a treatment dialogue. This method can be applied to citizen dialogue in the sense that experts with different views talk to each other honestly and with mutual respect and trust, responding to the contents of citizen dialogue and sitting next to citizens who conduct a dialogue in a small group. Multiplicity and a multi-layered structure of reality are gradually revealed by the experts' reflections or dialogue even if some participants have extreme views and believe that there is always only one truth (e.g., assessment of the risk of low-level radioactivity). The psychiatrist mentioned above argues that this approach may facilitate pluralistic views and generate third options that differ from pro- and anti-nuclear options.

6.3.2 Field Study Design

To conduct a case study on nurturing a culture of dialogue in Japan and applying a newly developed method for citizen dialogue, the research involved organizing and conducting a random sampling-based citizen dialogue on an environmental and energy policy in Kasugai City (population of approximately 300,000), Aichi Prefecture, or Omaezaki City (population of around 30,000), Shizuoka Prefecture, in 2015, 2016, 2017, 2018, and 2019. The citizen dialogue was on the disposal of high-level radioactive waste generated from nuclear power plants. Kasugai City belongs to the greater Nagoya region, the third-largest urban area in Japan. Omaezaki City is a host city for nuclear power plants, which were shut down immediately after the 2011 Fukushima accident at the request of the then Prime Minister. The dialogues were conducted in Kasugai City in 2015, 2016, and 2017 and in Omaezaki City in 2018. The dialogue in 2019 was held by connecting the meeting rooms in Kasugai City and those in Omaezaki City via the Internet (see the photo below). Fifty-three citizens participated in five dialogue events.

Citizen dialogue in 2019

Omaezaki City   Kasugai City

Note: Two venues in different cities were connected via the Internet. Extra rooms were used for small-group dialogue.

6.3.3 Results and Conclusion

The statistical (quantitative) analysis of the data obtained (attitudes for dialogue were measured empirically before and at the end of dialogue events), as well as qualitative content (theme) analysis of dialogue records among citizens, was conducted (Nakamura et al., 2021). The measure of attitudes toward dialogue was a five-point scale involving both listening and talking to others who held different views and positions. The five-point scales of policy preference options over four issues and a four-level scale on the participants' policy preference confidence were also reported. The latter measured changes in confidence regarding the preference level for a particular policy option.

The analysis revealed that the three proposed methods of dialogue—politeness-based facilitation dialogue, evidence-based and position-explicit presentations by experts with differing views and experts reflecting in tandem with citizens engaged in dialogue—might lead to enhanced positive attitudes toward dialogue with others holding different views, as well as better internal self-deliberation. The self-reporting ability to listen and talk to others who hold different views and positions increased through participation in citizen dialogue events. For example, 28 out of 53 participants answered "possible" to listening to others with differing views before the dialogue, while 38 out of 53 reported "possible" after the dialogue ($p = 0.045$ for the $\chi 2$ test of independence). The confidence level of the four policy preferences of contesting issues either increased, maintained, or decreased, depending on the issues at stake. For instance, 2 out of 36 respondents chose "very confident" regarding their assessment for long-term safety of geological disposals of high-level radioactive waste before the dialogue, while 10 out of 36 respondents selected "very confident" on their choice after the dialogue ($p = 0.011$). Binary logit regression analysis showed that reflection by experts in front of citizens was associated with a reduced confidence level after dialogue in two out of four policy issues, controlling the confidence level before dialogue. The latter indicator was used to measure the degree of internal deliberation and conviction on each participant's preference and righteousness for a designated topic. The case study suggested that explicit treatment of pluralistic positions and beliefs among citizens and experts is a primary enabler for quality communication to manage complexity and uncertainty.

6.4 Inclusive Partnership: A Vision for the SDGs

As shown in the two above cases, in particular the citizen dialogue case, inclusive partnership envisioned in goal 17 of the SDGs can be applied to a policy process in the following ways: (1) citizen participation is empowered and encouraged through the realization of a culture of dialogue, and (2) value-explicit communication and self-expression are secured by experts with different views. Since citizens are the only foundation of self-governance of communities and society, exclusion of citizens means abandonment of self-governance. Thus, the inclusion of citizens is a prerequisite for inclusive policy processes and shared governance (cf. Chambers, 1983). Moreover, putting value-explicit communication and self-expression by experts in second place is another important yet much more difficult step towards materializing inclusive governance (cf. Chambers, 1997). These two visions and practices advocated in the international development field are still significant and relevant in the larger context of SDGs, for pursuing higher level of governance.

In addition, putting the last first and putting the first last does not mean devaluing scientific reason. Instead, both reason and values/emotions should be appreciated and treated more carefully in a policy process to reduce coercion and domination. This means encouraging and incorporating the diversity and plurality of scientific reasoning, as well as the subjective values/emotions associated with personal judgment. Figure 6.1 shows a schematic diagram of the degree of respecting emotions/values and reasons in a policy process (horizontal axis) and the degree of supporting autonomy and spontaneity in a policy process (vertical axis). The explanation is as follows.

Regarding the left-side of the horizontal axis in Fig. 6.1, dealing with emotions and value systems in a policy process is an aspect that has received less attention. However, various ways to understand the roles of these elements in communication and self-/interpersonal management have been proposed and practiced in education, psychotherapy, spirituality, and conflict resolution. They include the art of awareness and emotional management, such as the Alexander technique, sensory awareness, mindfulness meditation, compassion, social and emotional learning (SEL), bioenergetics, and nonviolent communication (NVC), as well as transcend method for conflict transformation (Alexander, 1932; Brooks, 1974; Elias et al., 1997; Galtung, 2004; Halifax, 2018; Hanh, 1975; Huxley, 2010; Lowen, 1975; Rosenberg, 1999). The art of managing the collective creation of the future—beginning with escaping from past patterns of cognition and behavior and seeing and sensing the emerging future—also focuses on both the body and mind (Scharmer, 2016). The case studies introduced in this chapter, especially the citizen dialogue, tried to develop interlinkages between body and reason in a local policy process.

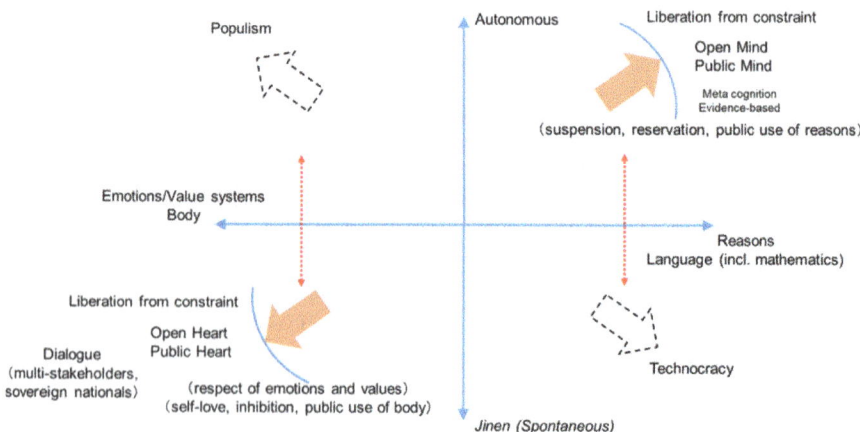

Fig. 6.1 Quadrants considering a less coercive policy process

The right side of the horizontal axis in Fig. 6.1 indicates the degree of reasoning in a policy process. This means further appreciation and inquiry of possible policy options and alternatives based on sound logic, data, and evidence. It requires the open-mindedness of researchers and experts, as well as stakeholders and citizens so that they do not adhere to the existing familiar views and ideas that they prefer. Intentional examination of existing ideas and unconscious mental models is needed, particularly when searching for missing links and new ideas.

The vertical axis of Fig. 6.1 encompasses the inclusiveness of autonomy and spontaneity (*jinen* in Japanese), both of which are requirements for less-coercive relationships within and between us. Autonomy is used in the sense of proactive self-governance in social networks and political actions. In this context, autonomy has a different meaning than when it is used in expressions like "the autonomic nervous system" (in the bodies of humans and other animals). The sense of being autonomic is not shown in Fig. 6.1. Such fluency resembles passiveness and non-intentionality in spontaneity. To realize a less-coercive relationship and state of being among human beings, appreciation and respect of both autonomy (say no to something/somebody, when it is violating your human dignity) and spontaneity (leave unnecessary intention, listen to your body and follow your authentic sense) would be needed. Making it easier is a challenge, especially in a policy process, which involves collective belief construction and actions. However, I would argue that it is no less important and difficult to reduce coercion in collective cases than in individual cases.

6.5   Concluding Perspective and Need for Further Experimental Research and Practice on the Ground

In the volcanic disaster management case study, the researchers acted not only as resource persons to contribute scientific knowledge but also as facilitators, jointly developing and maintaining the field of open dialogue, in collaboration with other stakeholders such as residents/citizens, local and national government officials, businesses, and media professionals. Consequently, the walls and borders between researchers and practitioners were intended to decrease. Researchers and practitioners' perceived roles and relationships were intentionally observed and experimented to check for possible changes toward better interlinkages and partnerships.

In open dialogue, the designed and intended sharing of pluralistic views is important to avoid falling into monologues or repetition of established statements. Securing openness and not-knowingness is necessary for both scientific knowledge and inference, as well as for individual views, values, and emotions. A scientific approach includes openness to alternative knowledge (hypotheses, models, and theories) that could explain data and phenomena better than existing knowledge. Moreover, multiple views, associated values, and emotions are allowed to ensure an external and internal dialogue between differences. This design of the field of dialogue may allow the creation of knowledge that renews participants' views with conviction. Such renewal of recognition about what we can do and who we are can be scientifically conducted, following the conventional practice of science, i.e., continual cycles of observation of the study target, hypothesis development, experiment to test the hypothesis, and modification of the hypothesis. Thus, the development and maintenance of open dialogue fields can be regarded as a scientific intervention by members of society to ensure better policymaking processes. This helps reduce the degree of domination and coercion, both for scientific content and views, as well as for interpersonal relationships and communication.

To materialize a safe and comfortable time and space for open dialogue, we must recognize three potential barriers and weaknesses in the body and mind. The first is the border and membership recognition of collaboration. In other words, with whom should I or do I want to collaborate to achieve a shared vision? Without a clear sense of sharing mutual visions, we would not collaborate with others. Thus, a border or wall of collaboration emerges. To realize the goals and targets of SDGs—all of which are individual, local, and global at one time in nature, we need to renew our perception of "with whom I would want to collaborate?" The concept of friends and enemies hinders the revision of our views. Sharing global and local goals, at least the goal of equal partnership (SDG 17; ends as means and means as ends), is needed.

The second potential barrier is a position that was explicitly examined and managed in the volcano disaster case study. Intentional matching and mixing of persons, positions, and organizations who rarely jointly work or communicate for dialogue is effective for the renewal of participants' beliefs without solid evidence and experience. Moreover, the intended design of randomness to develop a mutual dialogue among stakeholders is a plus to nurture open dialogue. This leads to fresh findings and learning. In the context of natural and social disasters and risk management, hazard as chance is literally an opportunity and coincidence, simultaneously.

The third and last is the biggest challenge: fear. Fear hinders honest and committed self-expression among dialogue participants. This makes it impossible to express weaknesses (or mistakes and errors in some cases). None of the participants disclose honest views and experiences even when they are free from positionalities. Also, some professionals, such as state persons and military officials, tend to demand suppression of weaknesses as professional norms and expectations. Thus, we cannot realize open dialogue when individual participants feel that they cannot and should not disclose their weaknesses. What can we do about this barrier?

Nearly all of us would feel afraid when we sense that we have to change views and beliefs that are more or less a basis of daily conduct. If such requests come from others as a force external to us, we naturally hesitate to change. Such perceived forced change would not actually occur, even when we understand why we have to change based on logical reasoning and/or a shared sense of norms and ethics. We also know that such forced change is not sustainable, without embodied conviction, even if it occurs. Interestingly, however, we also know that we can change when it is our authentic desire to change. We just do not want to change when it feels coercive. Without a sense of coercion, we can even enjoy the process of change. Securing a safe environment that allows the disclosure of weaknesses is difficult. Nonetheless, the politeness-based facilitation dialogue prescribed and tested in the case study of radioactive waste could be a starting point to explore how to realize a field of open dialogue.

Given the difficulty and significance of our weaknesses, further remarks are appropriate in the context of a risk-relevant policy process. Rules and regulations that allow researchers and government officials to disclose socially undesirable decision-making and execution processes after a disaster should be developed and embedded for better society-wide learning. Of course, we must not allow them to make mistakes beforehand. However, we also need to forgive them when they have erred and let them confess honestly after the disaster so that society learns from such mistakes. To do so, all sovereign nationals/citizens of Earth will make their own determinations based on their corresponding responsibility and capability. That is the meaning of sharing the vision and responsibility as a member of the human community (as well as a part of life on Earth) in the era of SDGs.

Acknowledgements I thank all individuals who participated and enabled the dialogical meetings in two case studies. This work was supported by (1) the Research Project to Support Regional Disaster Management of Ministry of Education, Culture, Sports, Science and Technology, Japan, (2) the Collaboration Research Program of IDEAS, Chubu University (IDEAS201505, 201605, 201706, 201806, and 201906), and (3) the Japanese Grants-in-Aid for Scientific Research (JP15K00656).

References

Alexander, F. M. (1932). The use of the self. *British Medical Journal, 1*(3728), 1149.
Brooks, C. V. (1974). *Sensory awareness*. The Viking Press.
Chambers, R. (1983). *Rural development-putting the last first*. Longman Scientific and Technical.
Chambers, R., & Cleaver, F. (1997). Whose reality counts? Putting the first last. *Project Appraisal, 12*(2), 134.
Elias, M. J., Zins, J. E., Weissberg, R. P., Frey, K. S., Greenberg, M. T., Haynes, N. M., Kessler, R., Schwab-Stone, M. E., & Shriver, T. P. (1997). *Promoting social and emotional learning: Guidelines for educators*. Ascd.
Galtung, J. (2004). *Transform and transcend*. Pluto Press, London.
Halifax, J. (2018). *Standing at the edge: Finding freedom where fear and courage meet*. Flatiron Books.
Hanh, T. N. (1975). *The miracle of mindfulness: A manual of meditation* (revised). Beacon Press.
Huxley, A. (2010). *The doors of perception: And heaven and hell*. Random House.
Kato, H., Shimizu, H., Kawamura, N., Hirano, Y., Tashiro, T., Yamashita, H., Tomita, K., Tomiyoshi, M., & Hagihara, K. (2014). A prospect toward establishment of basic and clinical environmental studies by ORT (On-Site Research Training). In *Basic and Clinical Environmental Approaches in Landscape Planning*, pp. 133–143. Springer.
Lowen, A. (1975). *Bioenergetics*. Penguin Books.
Matsumoto, M. (2012). *Kozo sai: Kagaku gijutu shakai ni hisomu kiki* (Structural disaster: crisis hidden in science and technology-based society). Iwanami Shoten.
Nakamura, H., Yamaoka, K., Horii, M., & Miyamae, R. (2019). An open dialogue approach to volcano disaster resilience and governance: Action research in Japan in the aftermath of the Mt. Ontake Eruption. *Journal of Disaster Research, 14*(5), 829–842.

Nakamura, H., Ueno, F., Higashihara, H., Hayashi, M., Sugita, S., & Fukui, H. (2021). Toward citizen dialogue-led environmental governance: An exploratory case study in Post-Fukushima Japan. *Environmental Management, 67*(5), 868–885.

Putman, N., & Martindale, B. (2022). *Open dialogue for psychosis: Organising mental health services to prioritise dialogue, relationship and meaning.* Routledge.

Rawls, J. (1999). *A theory of justice* (Revised). Belknap Press.

Rosenberg, M. B. (1999). *Nonviolent communication: A language of compassion.* Del Mar.

Saito, T., & Kainuma, H. (2016). Hairo wo kataru kotoba (Language to speak about decommissioning). In H. Kainuma (Ed.), *Fukushima Daiichi genpatsu hairo zukan (Encyclopedia of the "1F": A guide to the decommissioning of the Fukushima Daiichi nuclear power station)* (pp. 367–381). Ota Shuppan.

Scharmer, C. O. (2016). *Theory U: Leading from the future as it emerges* (2nd ed.). Berrett-Koehler Publishers.

Seikkula, J., & Arnkil, T. E. (2006). *Dialogical meetings in social networks.* Karnac Books.

Hidenori Nakamura is an associate professor of environmental policy at Department of Environmental and Civil Engineering, Toyama Prefectural University, Japan. He is also an AF-JSPS Postdoctoral Fellow at the University of Helsinki, Finland, during October 2021 to September 2022, where he studies a culture of open dialogue and sustainable development policy process in Finland. His interests include environmental governance, in particular citizen participation and a culture of dialogue. He has conducted a series of random sampling-based citizen dialogue on environmental and energy policy in Japan since 2015.

Part IV
Educational Perspectives

Chapter 7
Local Community-Based Education for Sustainable Development (ESD) During the COVID-19 Pandemic in the Asia and Pacific Region

Fumiko Noguchi

Abstract Given the systemic challenges Including the COVID-19 pandemic (see Chapter 1), the way of approaching and thinking to know and understand the problem seems neither to give a perfect answer nor a clear direction to follow to the emerging sustainability challenges that the world faces now. This is because the causes and impacts of a single problem are complexly connected to the causes and impacts of other problems. The adaptive approach seems to be effective and appropriate to these wicked problems. With this, diverse stakeholders participate in finding a solution by bringing their knowledge and experiences. This is an educative process that can be called Education for Sustainable Development (ESD) in a community context. Stakeholders mutually learn and co-create new knowledge, through dialogues, collaborations and even through a clash of tensions between different interests and positionality. This chapter illustrates how the adaptive approach, or co-learning and cocreation of knowledge for sustainability, took place during the COVID-19 pandemic, drawing the experience of the Regional Centre for Expertise on ESD (RCEs) in the Asia and Pacific region by United Nations University. It discusses the factors that promote or hinder the activities and supporting mechanisms are examined. In particular, it highlights the issue of 'digitalisation' that was emerged in the community-based ESD efforts that brought an opportunity and a challenge to multistakeholder participation.

7.1 Introduction: Adaptive Approach and Education for Sustainable Development (ESD)

Whether it is the Sustainable Development Goals (SDGs) (UNGA, 2015) or its central concept of sustainable development, what is agreed in the international policy framework can provide only the broad direction. In reality, the only way to achieve sustainable development is to think about and take an action for what we can do

F. Noguchi (✉)
United Nations University Institute for the Advanced Study of Sustainability (UNU-IAS),
Tokyo, Japan

in our everyday life context at local community levels (Fien & Tilbury, 2002). The problems we face in our everyday life are complex and causal and often go beyond the understanding of existing political and theoretical discourses.

The COVID-19 pandemic has given a major impact on many aspects of our life. The existing disparities have widened further, and the socially and economically vulnerable are placed in more difficult circumstances (United Nations, 2020), facing unequal access to social service and having psychological instability, lower incomes, fewer employment opportunities and fewer educational opportunities. At the same time, the pandemic triggered a change in our ways of life. The mobility restriction has also increased us to localise by opting for online working modes (Haski-Leventhal, 2020). This trend urges some companies and people to choose to re-settle in rural communities, rather than in central urban capitals. In this current complex and uncertain situation, a question emerges: "Should we rebuild a pre-COVID-19 society?"; or "Should we transform our society, with emphasis on choosing the ways in which we can be more resilient and sustainable than before? (Carr, 2020; OECD, 2020).

Climate change-related disasters, expanding socio-economic gaps and the COVID-19 pandemic—with those emerging sustainability challenges that the world faces now, it is getting clearer than what we have thought as certain, including the way of approaching and thinking to know and understand the problem seems neither giving a perfect answer nor a clear direction to follow. What has become clear is that the causes and impacts of a single wicked problem are complexly connected to the causes and impacts of other problems. As such, we are facing systemic challenges (see Chapter 1). Many of us, including researchers, policymakers and NGOs—those who are called 'experts', have to question the majour premises in our familiar syllogisms for the problem, upon which we develop policy, research and practice for the solution.

The adaptive approach can provide a clue in finding an effective solution for those emerging challenges. This is the approach that has been discussed in the area of environmental management, which aims to find the solution effective and appropriate to the local community issue by bringing knowledge and experiences of diverse stakeholders (Kingsford et al., 2017). And this learning takes place all the management cycle including "framing questions and problems, undertaking experimentation and testing, critically processing the results, and reassessing the policy context that originally triggered the investigation in light of the newly acquired knowledge" (Stankey et al., 2005, p. 7). As we all know well, this process can contain the tensions and issues of power imbalance amongst the stakeholders. Learning occurs even during the efforts to overcome those uncomfortable challenges (Jacobson et al., 2009).

This process of co-learning and co-creation of knowledge in an adaptive approach overlaps with the core philosophy of Education for Sustainable Development (ESD), which is addressed in SDG Target 4.7. Education in Target 4.7 ensures inclusive and equitable opportunities to access quality education and lifelong learning for all through formal, non-formal, and informal education and learning (UNESCO, 2019). Education in ESD goes beyond formal education. ESD needs to be implemented and promoted through a lifelong process. This point is also emphasised in the global policy framework to support ESD promotion by UNESCO. While the role of ESD in the SDGs is considered just as one of the 17 goals or 169 targets, it holds the key to achieving all the SDGs. To this point, "ESD for 2030" framework, which was launched in 2020 at the initiative of UNESCO, reassures the critical role of Target 4.7 in linking different sectors and fields that critically interrogates the tensions and ambiguities between the different goals and contextualising them. ESD lies in the process of learning and participation of diverse stakeholders from different fields and sectors to share their experiences and knowledge and act together for transforming themselves and their surrounding situation for the better, by over-coming conflicts, authority structures and power issues associated with knowledge and positions (Noguchi, 2016, 2022).

This chapter reports findings of the research project, "Rethinking Community-based ESD in the COVID-19 Era", conducted by the United Nations University, Institute for the Advanced Study of Sustainability (UNU-IAS) in 2020. This research examined how the adaptive approach, or co-learning and co-creation of knowledge for sustainability, took place during the COVID-19 pandemic, drawing the experience of the Regional Centre for Expertise on ESD (RCEs) in the Asia and Pacific region. This research investigated the effectiveness and relevance of RCEs' community-based and multi-stakeholder partnership to the current local community situation under the pandemic. It identified the factors that promote or hinder the activities and supporting mechanisms such as funding, policy, networking, etc. In particular, this chapter highlights 'digitalisation' that brought an opportunity and a challenge to the community-based ESD. The experience of RCEs can provide an insight to understand how the core element of the adaptive approach, or co-learning and co-creation of knowledge, takes place in this circumstance of the COVID-19 pandemic.

7.2 RCE Research Project: "Rethinking Community-Based ESD in the COVID-19 Era"

RCEs are the networking project convened by the UNU-IAS since 2003, which was launched in 2003 in response to the 2002 United Nations General Assembly resolution on the Decade of Education for Sustainable Development (UNDESD) (UNESCO, 2022a). It takes four approaches to its activities, including (1) local community-based; (2) multi-stakeholder partnerships; (3) learning and education centred; and (4) making a linkage between and coordinating with relevant international, national and local policies on sustainable development and ESD (UNU-IAS, 2014). The key approaches of RCEs overlap with the key credentials of the adaptive approach (UNU-IAS, 2014).

UNU-IAS serves as the Secretariat of the Global RCE Network. It brings together the individual RCE, which is also the local network of stakeholders such as universities, NGOs, local governments and businesses/industries who collaboratively work for the solution of sustainability challenges in local communities with the focus on education as a key theme. As of 2021, 181 RCEs are acknowledged in the world

(UNU-IAS, 2021). During the five years of the Global Action Programme (GAP) on ESD (2015–2019) (UNESCO, 2022b), the UNU-IAS counted 479 projects as implemented ones, 67% of which were conducted through multi-stakeholder partnerships (Vaughter & Noguchi, 2020). RCEs also have given the inputs to global and national policy frameworks, based on their local community experiences, including the UN Decade of Education for Sustainable Development (UNDESD: 2005–2014), GAP and ESD for 2030, and SDGs (UNESCO, 2021a, 2021b). Those local RCEs also share the knowledge and experiences within respective regional networks in the Asia and Pacific, Africa and the Middle East, North and South America, and Europe, where there are independent governing bodies.

The Asia–Pacific region was selected for four reasons as below.

a. There are more experienced RCE members in this region which engaged with ESD activities longer than the other three regions;
b. There were the largest number of RCEs amongst the four regions; 67 in Asia Pacific, 44 in Europe, 38 in Africa and the Middle East, and 26 in the Americas) as of 2020;
c. Unlike the other three regions, there is the RCE Asia–Pacific Regional Coordinating Committee (APCC), which is governance functioning for decision-making, liaison and coordination of activities and strategy development in the region; and,
d. I have already established a relationship of trust with the APCC

As such, the data was collected and analysed based on the participant observation at a series of public webinars hosted by RCEs in the Asia Pacific region and distribution of questionnaires to all the RCEs in the region and focus groups with RCEs in Japan (See Table 7.1).

Table 7.1 Process of data collection

Situational analysis (May–July 2020)
– Desk Review
– Participant observation at three public webinars organised by RCE Srinagar (India)   • 1st session: 18th May, 14:00–17:00 (JST), presented by 16RCEs, about 100 viewers   • 2nd session: 9 June, 14:00–16:00 (JST), presented by 9RCEs, about 65 viewers   • 3rd presentation: 11 June, 14:00–16:00 (JST), presented by 12RCE, about 50 viewers
– Unstructured Interviews with selected RCEs
– Questionnaire development with the Regional Coordinating Committee for Asia Pacific RCEs
Data collection and analysis (August-November 2020)
– Questionnaire was distributed to all RCE offices in the Asia Pacific region (17 RCEs responded)
– Participant observation at the 13th RCE Regional Meeting Webinar Series where 16 respondent RCEs presented their responses (24 September)
Focus group with 8 RCEs in Japan (24 August and 18 November)

7.2.1 Participant Observation at Webinar Series

In May and June 2020, a series of three webinars on "ESD Action in COVID-19 Era: Experiences of RCEs of Asia–Pacific Region" was co-organised by RCE Srinagar and RCE Greater Western Sydney on 14 May, 9 June and 11 June. This shared how RCEs were responding to the pandemic through their activities by the presentations of 37 RCEs, 55% of the Asia Pacific Region (UNU-IAS, 2020b). These presentations demonstrated that 26 RCEs (70%) did not stop their activities by continuing or modifying their planned activities. 13 RCES (35%) started new projects based on the needs analysis of local situations. The presentations highlighted the emerging challenges in local communities and the diverse types of efforts made by RCEs, which covered the themes of health, employment, food security, health and educational continuity (WSU&RCE Srinagar, 2021). For example, RCE Greater Western Sydney (Australia) developed and shared e-learning resources for home education through close collaboration with environmental education centres and zoos. RCE Kitakyushu (Japan) created an SDGs library where the local residents could borrow the books on ESD and SDGs via postal service. RCE Greater Gombak (Malaysia) conducted Need-Based Research, upon which they conducted the project to deliver the essential items and food to the indigenous and refugee communities. RCE Sundarbans (Bangladesh) conducted a series of webinars on academics-practitioners dialogue on the theme of poverty, Rohingya refugees and microcredit. RCE Chandigarh (India) created e-learning resources and provided online courses on energy efficiency for brick kiln owners and firemen. RCE Yogyakarta (Indonesia) worked with batik clothing manufacturers to produce personal protective equipment such as masks, creating job opportunities for local women.

There were noticeable activities in which youth actively participated. The youth group of RCE Yokohama (Japan) broadcasted the community efforts that tackled the socioeconomic challenges during the pandemic. RCE Greater Phnom Penh (Cambodia) changed their organic farming training to the farmers in rural communities from in-person to online due to the travel restrictions. In order to make learning effective, they closely worked with local youths using their smartphones to support the learners who were not familiar with ICT. These cases show that the youth members are actively involved in the planning of a project to help them to develop a sense of ownership as a member of the local community.

The issue of 'digitalisation' emerged through the presentations. 14 RCEs (37%) changed the mode for ESD activities to online. Digitalised activities included e-resource development, e-learning programme provisions and organisation of online conferences, symposiums and workshops. The change to the online mode was accompanied by the challenges in the investigation of the effective teaching methodology (RCE Kyrgyzstan), and securing access to online learnings in the communities with limited infrastructures and lack of ICT devices and literacy (RCE Kodagu and RCE Okayama).

7.2.2   Data Collection Through Questionnaire Distribution

By linking to the planning process for the four online sessions of the 13th RCE Regional Meeting in September–October 2020, the questionnaire was jointly prepared with RCE Kyrgyzstan, which hosted the meeting, the Regional Advisor for Asia and the APCC and distributed to all RCE focal points in the Asia Pacific region in July 2020. The questionnaire consisted of four questions asking; (i) Impact of the pandemic to the local community; (ii) Impact of the pandemic on the activities of the RCE; (iii) The role of RCE's in responding to the pandemic including emerging challenges and opportunities; (iv) Supportive measures required. 17 RCEs from 10 countries responded (25% response rate). These respondents also made presentations based on their responses to the questionnaire at the 13th RCE Regional Meeting (UNU-IAS, 2020a). The following is a summary of the main responses.

7.2.2.1   Impact of the Pandemic on Local Communities

The responses to the questionnaire and the presentations at the online regional meetings indicated that local communities faced a wide range of regional issues, including restrictions on imports and exports due to border closures, closure of markets, city lockdowns, lack of health care systems, digitalisation in businesses and educational activities; loss of teaching and learning, widening of economic, social and educational gaps. Amongst responses, the impact of digitalisation on their local activities is noticeable. While some are saying that digitalisation increased networking and learning opportunities, some groups of people and areas of activities were left behind in digitalised activities (See further discussions in 2.4).

7.2.2.2   Impact of the Pandemic on RCE Activities

The responses to the questionnaire showed that the pandemic forced all RCEs to cancel or change their activities, due to the closure of activity centres and schools, restrictions on movement, social distance and face-to-face activities and the suspension of grants. Experience-based and practice-oriented activities, such as nature-based outdoor education and agricultural training were often cancelled.

7.2.2.3   RCE's Response to the Pandemic

The responses showed that all the RCEs sought a way to respond to the local challenges through interviews, questionnaires and participatory observation and the digitisation was opted by 15 RCEs (88%) as the majour tool to respond to the situation. The activities carried out through digitalisation included lecture-based training, workshops, meetings, and information sharing. These also include the development of online training programmes on infection prevention and online environmental education materials for home learning.

The 2 RCEs' responses characterised the change in the way of the partnerships amongst diverse stakeholders during the pandemic. The response from India suggests that they newly developed partnerships with the health sector of the government with whom they had no previous contact, in order to start the activity to provide hygiene guidance to local communities and schools. Cambodian case indicates that they developed new partnerships with rural administrators and experts or experienced farmers in remote rural areas, instead of sending experts from centres such as big cities to remote rural areas, due to mobility restrictions.

8 RCEs (47%) in Malaysia, India, the Philippines, Bangladesh, Indonesia and South Korea conducted activities that targeted vulnerable groups in the community, such as ethnic minorities, indigenous peoples, refugees, migrants and single parents. These included the interviews and supporting activities, such as distributing relief materials, job creation such as making masks and protective clothing.

A response from Japan also indicated that the pandemic complicated the implementation of activities that urgently respond to the local community problems such as relief work at the natural disaster-affected areas during the pandemic, facing the challenges such as mobility restrictions and infection prevention measurement, and lack of clear policy measurement.

7.2.2.4 Supporting Measures Required

The questionnaire responses suggested the following could facilitate and support the community-based ESD activities during the pandemic:

- Clear government priorities and guidelines for the implementation of community-based ESD activities during the pandemic;
- Funding for the implementation of activities
- Capacity building opportunities and emotional support for local ESD practitioners
- Networking opportunities for sharing the relevant information and local practice and policy advocacies to promote community-based ESD during the pandemic.

7.2.3 Focus Groups for RCEs in Japan

In August and November 2020, online focus groups were conducted with 8 RCEs in Japan. These aimed to gain a contextual understanding on the data collected from the participant observation at the webinar series and the questionnaire. In line with the content of the questionnaire, the following three themes were discussed in plenary and breakout sessions:

- Implications of COVID-19 for regional and RCE activities and main challenges;
- Drivers and challenges to promote activities; and,
- Short- and long-term strategic directions.

Two focus groups identified the role that RCEs played in assessing the impact of the pandemic on local communities and seeking the appropriate way best to respond to the challenges that local communities face. Like the RCEs in other countries in the region, they cancelled the planned activities and conducted new activities to respond to the situation flexibly. The focus groups identified the technological digital divide, in which some community members, such as the elderly and socially and economically vulnerable groups were left-behind, having little access to online information and social services, because of their little ICT skills and knowledge and lack of ICT infrastructure in rural areas. 'Tacit digital divide' also emerged from the focus groups. Some RCEs identified some youth members who were familiar with ICTs and knew the online networking events isolated feeling loneliness. Other RCEs shared the challenges that they experienced with local governments that were not able to keep up with the digitalisation. This also interrupted in providing social services to the local community members and switching to online education at local schools. While networks, conferences, seminars, and local information dissemination seemed to be relatively easy to be provided online, the RCEs also identified that there were areas of activities that had to be conducted in on-site and face-to-face mode, such as support for small children and elderly, people with intellectual challenges and mental illness, and emergency relief works in disaster-affected areas.

The discussions highlighted the RCEs' experience of more active youth engagement with the community-based ESD activities compared to pre-pandemic. Several RCEs observed the proactive participation of youth in designing and implementing the local community activities using ICT tools. For example, the university student group co-designed and co-conducted the online environmental education programme with school teachers (RCE Greater Sendai). The youth group organised the networking meetings for the 1st year university students and the seminars to discuss the community-based volunteer activities under the pandemic (RCE Hyogo-Kobe). Those youth planned, covered the local community activities, disseminated through social media and provided the learning support of the school children, through an equal partnership with the diverse adult stakeholders. Above all, youth groups quickly responded to the change to digitalisation and conducted online mode activities. In this process, it should be noted that the youth group of RCE Hyogo-Kobe discussed identifying what activities should/could not be digitalised in their local community context. This indicates the change of the position of youths in community-based ESD activities where the adults often plan and assign the role of youths in local community projects.

7.2.4 Digitalisation

The above discussed the current status of community-based ESD drawing on the data gathered from the webinars, questionnaires and focus groups with the RCEs in Asia and the Pacific region. The research emerged the issue of 'digitalisation' that brought both opportunity and challenge to the community-based ESD activities in the region.

The challenge involves two types of 'digital divide' that obstructed the community-based ESD; 'technological' digital divide and 'tacit' digital divide. The data gathered at the webinars and through the questionnaire revealed the 'technological' digital divide, which involves the disparity of information and access to communication in the region. This includes the issues related to more about the lack of resources, infrastructure and ICT equipment to support the internet, financial resources in accessing these, and ICT literacy, as well as difficulties in implementing practices in the area where digitalisation was not physically possible, particularly in logistics, medical and material support activities. Lack of infrastructure and equipment did not keep up with the necessity of digitalization in the activities in developing countries in the Asia–Pacific region and in rural areas of Japan since the pandemic.

Additionally, the issue of the 'tacit' digital divide occurred in a context where access to infrastructure and ICT equipment is available, and where a certain degree of IT literacy is expected. The research identified the problem of isolation of youth members amidst a number of online networking opportunities, and that there were learning activities that were not fully and effectively passed through digital devices or online-mode. Some RCEs faced challenges in conducting online-mode nature-based experiential learning activities in a local community context, such as outdoor and environmental education and farming. This problem poses a question about the limitation of digitalisation that may not appropriately address a particular type of emotion, knowledge and approach to learning. It is assumed the core of those activities, which are underpinned with local knowledge and the way of learning local knowledge, was built on a highly physical and sensory process of learning in a natural and local community setting. Local knowledge, including traditional and indigenous knowledge, is often acquired by using hands, body and sense tacitly through observation and by spending time and space with others. The relevance and effectiveness of digitalisation to local knowledge and the learning of local knowledge need to be further investigated.

The opportunity that digitalisation brought should be highlighted, in particular, the change in the quality of youth members' participation in the community-based ESD activities. Youth became more actively involved in deciding the direction and taking leadership in activities with diverse stakeholders in RCE Greater Phnom Penh, RCE Yokohama, RCE Greater Sendai, RCE Hyogo-Kobe. Active usage of ICT and social media enables them to establish a stronger position and to speak in their own voices so that they can develop their equal partnership with the other 'adult' actors in a local community. Unfortunately, there are many examples where local adults had already planned the activities and the youth have participated in them in a tokenism way. Further analysis needs to be conducted to know how degree and quality of empowerment are different between the one that comes from finding an active role and gaining confidence in a local community within a role prepared by the other (adult) actors (tokenism participation), and the one that is obtained through gaining confidence through actively engaging with the whole project cycle from designing to evaluation (full participation).

7.3   Discussion and Conclusion

The pandemic has accelerated digitalisation in every part of our life including education. Global communities and national governments have been concerned about the left behind in formal education. UNESCO points out that in primary and secondary schools alone 250 million children are out of school due to blockades and school closures. Of these, 23.8 million are at risk of not returning to school when schools reopen (Giannini, 2020). Responding to this situation, the policies to support the digitalisation of education is rapidly developed and implemented in many parts of the world. As seen in the cases of RCEs in the Asia Pacific Region, stakeholders in the areas of community-based ESD, including non-formal and informal education practitioners and policymakers, such as community learning centres and NGOs, business and local governments, have been investigating a new direction. To seek this direction, the stakeholders are paying attention to digitalisation as a key tool to solving complex local community challenges by empowering and building the capacity of local community members.

However, the experience of RCEs indicates that digitalisation seemed to work better with transmitting a particular type of knowledge and learning that was captured well with a digital tool. That could include cognitive, explanatory and generalised knowledge and learning, wherein modern scientific knowledge and learning could stand out compared to traditional and indigenous knowledge and learning that could be tacit, non-cognitive and socially and environmentally contextual. Digitalisation has brought opportunities, in that it allowed more diverse stakeholders who never had chances before to engage with community-based ESD as long as the accessibility is assured. On the other hand, the above research findings also imply that digitalisation may automatically filter knowledge and learning that are suitable for digital-based information transmission, making others unnoticeable. Those 'others', such as local, practical, indigenous, and embodied knowledge, may be able to provide essential knowledge and learning that should be integrated into the efforts for solving the local community challenges. Rather than rushing to digitalisation, as the aforementioned experience of RCE Hyogo-Kobe tells, it is important to critically examine how and why digitalisation can help empower and build the capacity of stakeholders in their efforts for actions for sustainability and to develop projects and relevant policies, without leaving behind local knowledge and learning and the local knowledge holders.

Key elements of the adaptive approach, which were presented in the introduction, were observed in the RCEs' efforts for making the local communities resilient and sustainable during the COVID-19 pandemic above discussed. It could be argued that, in a very changing situation where there was no clear policy guidance and scientific projection on the trajectory of COVID-19 available, they just took an adaptive approach unintentionally as they may have had no choice but to bring knowledge, experiences and voices of diverse stakeholders, to find the appropriate solutions for existing and emerging sustainability challenges. This process created co-learning opportunities, which were informally embedded in the actions of the stakeholders in their attempt to find the appropriate solution to the local community issues in this uncertain time. In this process, the issue of digitalisation emerged. It brought benefits and challenges in terms of multistakeholder participation. It changed the way of collaboration and partnerships, which allowed more active and equal participation of diverse stakeholders, such as youths while creating technological and tacit divides in the local communities. The local communities rushed to use the digital technologies and the policies were facilitating the process. Here, it should be stressed that digitalisation is just a means, not a goal of the community-based ESD activities. How digitalisation has changed the way of co-creation of knowledge and co-learnings amongst the multi-stakeholders need to be further investigated. In particular, several key issues need to be critically analysed, such as what knowledge and learning of whom stand out or overlooked, how and what tensions occur, and what can facilitate inclusion and participation of diverse stakeholders when digitalising the activities. Through this investigation, the adaptive approach can be updated and more operational which can be applied to different educational communities to work on SDGs.

The new global policy framework, "Education for Sustainable Development: Towards achieving the SDGs (ESD for 2030)" was launched in 2020 by UNESCO, the leading agency for the promotion of ESD. This aims to contribute to achieving SDGs through education. The road map for ESD for 2030 was developed, outlining actions in five priority action areas on policy, learning environments, building capacities of educators, youth and local level action. Each UNESCO Member State is expected to develop an Action Plan inviting relevant stakeholders to five priority action areas from diverse sectors and fields (UENSCO, 2021a, 2021b). Local ESD practitioners should be one of them. In the process of developing international and national policies on the ESD for 2030 framework, building a resilient and sustainable society with and after the COVID-19 pandemic becomes an essential theme. The success of the SDGs relies on the successful development and implementation of ESD for 2030 policies. Ultimately, how to build back better relies on how local actors can participate in this platform and deliver realistic, relevant and effective policies with local communities, based on their experiences on community-based ESD practices. To this point, the RCEs experiences give insights on ESD at local communities and its potential opportunities and challenges arising in relation to digitalisation.

Acknowledgements To conduct a data collection of RCEs in the Asia Pacific region, the Asia Pacific RCE Coordinating Committee, the Asia Pacific RCE Advisors, the UNU-IAS ESD

Programme staff provided their extensive advice and support. I would like to thank all of them for their valuable contributions. I would also many RCEs in this region, who shared their experiences during the pandemic. Lastly, I would like to acknowledge that this study was carried out as part of the RCE project funded by the Ministry of the Environment, Japan. I would also like to thank their continual support for this global network.

References

Carr, P. (2020). Returning to 'normal' post-coronavirus would be inhuman. *The Conversation*. https://theconversation.com/returning-to-normal-post-coronavirus-would-be-inhumane-136558. Accessed 8 February 2022.

Fien, J., & Tilbury, D. (2002). The global challenge of sustainability. *Education and Sustainability: Responding to the Global Challenge, 1*.

Giannini, S. (2020). *School, health and nutrition: Why COVID-19 demands a rethink of education to address gender inequalities*. UNESCO. https://en.unesco.org/news/school-health-and-nutrition-why-covid-19-demands-rethink-education-address-gender-inequalities. Accessed 8 February 2022.

Haski-Leventhal, D. (2020). *The lighthouse: Seven positive outcomes of COVID-19*. The Lighthouse, Issue. https://lighthouse.mq.edu.au/article/april-2020/seven-positive-outcomes-of-covid-19. Accessed 27 October 2021.

Jacobson, C., Allen, W., Veltman, C., Ramsey, D., Forsyth, D., Nicol, S., Allen, R., Todd, C., & Barker, R. (2009). Collaborative learning as part of adaptive management of forests affected by deer. In C. A. G. Stankey (Ed.), *Adaptive environmental management: A practitioner's guide* (pp. 275–294). Springer Science + Business Media B.V.

Kingsford, R., Roux, D., McLoughlin, C., Conallin, J., & Norris, V. (2017). Strategic adaptive management (SAM) of intermittent rivers and ephemeral streams. In T. Datry, N. Bonada, & A. Boulton (Eds.), *Intermittent rivers and ephemeral streams* (pp. 535–562). Academic Press.

Noguchi, F. (2016). Radical approach from the periphery: Informal ESD through rights recovery for indigenous Ainu. In J. Singer, T. Gannon, F. Noguchi, & Y. Mochizuki (Eds.), *Educating for sustainability: Fostering resilient communities after the triple disaster* (pp. 201–215). Routledge.

Noguchi, F. (2022). *Rethinking Education for Sustaianable Development in a Local Community Context*. Springer.

OECD. (2020). *Building back better: A sustainable, resilient recovery after COVID-19*. Tackling coronavirus (COVID-19), Issue. https://www.oecd.org/coronavirus/policy-responses/building-back-better-a-sustainable-resilient-recovery-after-covid-19-52b869f5/. Accessed 8 February 2022.

Stankey, G. H. (2005). *Adaptive management of natural resources: Theory, concepts, and management institutions*. Department of Agriculture.

UNESCO. (2019). *Framework for the implementation of education for sustainable development (ESD) beyond 2019*. https://unesdoc.unesco.org/ark:/48223/pf0000370215. Accessed 8 February 2022.

UNESCO. (2021a). *Education for sustainable development: A roadmap*. UNESCO. https://unesdoc.unesco.org/ark:/48223/pf0000374802. Accessed 8 February 2022.

UNESCO. (2021b). *Global action programme on education for sustainable development (2015–2019): GAP partner networks*. UNESCO. https://en.unesco.org/gap/partner-networks. Accessed 27 October 2021b.

UNESCO. (2022a). *UN decade of ESD*. https://en.unesco.org/themes/education-sustainable-development/what-is-esd/un-decade-of-esd. Accessed 8 February 2022.

UNESCO. (2022b). *Global action programme on education for sustainable development (2015–2019)*. https://en.unesco.org/globalactionprogrammeoneducation. Accessed 8 February 2022.

UNGA. (2015). *Transforming our world: The 2030 agenda for sustainable development (A/RES/70/1).* https://www.un.org/ga/search/view_doc.asp?symbol=A/RES/70/1&Lang=E. Accessed 8 February 2022.

United Nations. (2020). *Shared responsibility, global solidarity: Responding to the socio-economic impacts of COVID-19.* United Nations. https://unsdg.un.org/sites/default/files/2020-03/SG-Rep ort-Socio-Economic-Impact-of-Covid19.pdf. Accessed 8 February 2022.

UNU-IAS. (2014). *Ten years of regional centres of expertise on education for sustainable development.* UNU-IAS. https://www.rcenetwork.org/portal/sites/default/files/public_resource/01_UNU_10yearsBook_web.pdf. Accessed 8 February 2022.

UNU-IAS. (2020a). *The 13th Asia-Pacific RCE regional meeting emphasises collaboration to build capacity for sustainability.* UNU-IAS. https://www.rcenetwork.org/portal/13th-asia-pacific-rce-regional-meeting-emphasises-collaboration-build-capacity-sustainability. Accessed 27 October 2021.

UNU-IAS. (2020b). *Webinar series 'ESD action in COVID-1 9 ERA: Experiences of RCEs of the Asia Pacific region' concluded.* UNU-IAS. https://www.rcenetwork.org/portal/webinar-series-esd-action-covid-19-era-experiences-rces-asia-pacific-region-concluded. Accessed 27 October 2021.

UNU-IAS. (2021). *RCE vision and mission.* UNU-IAS. https://www.rcenetwork.org/portal/rce-vision-and-mission. Accessed 27 October 2021.

Vaughter, P., & Noguchi, F. (2020). *RCE project trends during the global action programme on ESD.* UNU-IAS. https://rcenetwork.org/portal/sites/default/files/flipping_book/pdf/RCE_Pro ject%20Database%20insights_online.pdf. Accessed 8 February 2022.

WSU & RCE Srinagar. (2021). *Webinar series report: ESD action in COVID-19 era: Experiences of RCEs of Asia-Pacific region.* https://westernsydney.edu.au/__data/assets/pdf_file/0010/1749331/ESD_COVID_Report_Final.pdf. Accessed 8 February 2022.

Fumiko Noguchi is a Research Fellow at United Nations University Institute for the Advanced Study of Sustainability (UNU-IAS). With 25 years of work experience, she engaged with practices, policy advocacies, research, teaching and module development in the area of Education for Sustainable Development (ESD) with a focus on a local community-based and multistakeholder approach to ESD. In particular, she has investigated the epistemological power imbalance amongst diverse types of knowledge and learning approaches of diverse stakeholders and searched for a way to realise a real dialogue for sustainability amongst them. Her majour publications include Rethinking Education for Sustainable Development in a Local Community Context (Monograph, Springer, 2022).

Chapter 8
Shaping Sustainability Priorities for Higher Education Institutions

Vincent C. H. Tong

Abstract Higher education plays a vital role in providing intellectual and community leadership in addressing the United Nations' Sustainable Development Goals (SDGs). To achieve these goals at the institutional levels, universities need to define and articulate their strategic priorities for a wide range of stakeholders. University strategy documents form a key documentary source for investigating strategic positioning, including how institutional values, functions, and stakeholders are involved in shaping the institutional strategic priorities. In this contribution, I explore the framing of sustainability priorities in relation to the complex functions of higher education institutions—by analyzing eight institution-level SDGs-aligned strategies published by universities in different parts of the world. The analysis shows that there is generally a dichotomy between the strategic priorities in core university functions (i.e., research, education, and engagement) and those associated with campus operations. Based on the findings, I conclude that inclusive and diverse partnerships will help enhance the shaping of institutional sustainability priorities, thereby making universities truly resilient institutions to advance sustainability for the communities they serve.

8.1 Introduction

The United Nations' Sustainable Development Goals (SDGs) provide a framework for capturing and addressing a diverse range of complex and pressing sustainability issues that humanity is facing. The SDGs consist of 17 goals and 169 targets designed to "end poverty, protect the planet and ensure that by 2030 all people enjoy peace and prosperity" (United Nations Development Programme: https://www.undp.org/sustainable-development-goals). The higher education sector plays an active role in advancing these values-based strategic goals—through working in partnership with its extensive networks of stakeholders and communities for the global common good.

V. C. H. Tong (✉)
Department of Geography and Environmental Sciences, Northumbria University,
Newcastle upon Tyne, UK
e-mail: vincent.tong@northumbria.ac.uk

This approach itself also directly addresses Goal 17 for "strengthening the means of implementation and revitalize the global partnership for sustainable development."

Apart from initiating, leading, and taking part in community networks for advancing sustainability, universities and other higher education institutions also provide intellectual leadership for shaping our understanding of the SDGs themselves—not only within the domains of each sustainability goal but also the interlinkages across the targets. The depth and breadth of this academic leadership are perhaps best exemplified by the sustainability research published by academics from diverse disciplines and backgrounds. Scholars have, for instance, used the SDGs as the foundation to develop synergistic approaches for supporting sustainable development (e.g., Liu et al., 2018), and have investigated the interconnections between the social and natural factors in sustainable development (Reyers & Selig, 2020). The research on the linkages between climate actions and the other SDGs (Nerini et al., 2019) has significant practical implications on the development of synergies between SDGs-aligned initiatives and ways to maximize their effectiveness for addressing the climate crisis. The study on the disconnect between the SDGs (McGowan et al., 2018) provides important evidence and foundation for enhancing policymaking in sustainability. These research efforts clearly show the importance of higher education' roles in establishing the theoretical dimensions of the SDGs, with implications for shaping the policymaking process and enhancing the implementation of the SDGs across a wide spectrum of contexts beyond academia.

Many higher education institutions around the world have taken the strategic leadership to address the SDGs since its launch in 2015. The Times Higher Education (THE) Impact Ranking, which is an annual global performance table for assessing higher education institutions against the SDGs, clearly demonstrates the strategic significance of the United Nations' framework across the higher education sector. The launch of *Nature Sustainability*, a high-profile transdisciplinary academic journal with direct reference to the SDGs in its aims and scope (https://www.nature.com/nat sustain/aims), underscores the growing research and education interests in sustainability as a focal point of academic endeavors. These two examples with comprehensive links across all the SDGs indicate that sustainability has become a significant theme in the agenda across the higher education sector. Given higher education's deep involvement in advancing the SDGs, it is crucial for universities to prioritize their resources and strategize their approaches to sustainability.

University strategies provide documentary evidence for studying institutional strategic positioning in higher education (Fumasoli et al., 2020). They are therefore an important source for understanding the interface between the diverse sustainability goals and complex higher education functions—with a focus on the organizational perspectives that reflect the different institutional and cultural settings. As highlighted by Fumasoli and Lepori (2011), university strategies are about intentions and actions, which are closely related to organizational values, structures, and processes. The framing of strategic principles and objectives reflects the institutional identities and values underpinning how universities see themselves in connection with the communities they interact with and serve. It is not uncommon to see statements featuring core institutional values in the opening sections of university strategy documents.

Institutional strategies also show how universities identify emerging opportunities by building on existing institutional strengths and developing new directions. They indicate the extent to which universities value their partners, accountability, and transparency. Details about steering groups, implementation plans, and measurements of success are often found in institutional plans. SDGs-aligned strategies therefore allow us to develop a better understanding of the formal and informal institutional processes involving both university communities and their external partners in decentralized organizational cultures.

The aim of this chapter is to explore the shaping of sustainability priorities in relation to the complex functions of higher education institutions. More specifically, I will focus on how these functions are foregrounded and embedded in institution-level strategies that make direct references to the SDGs. Drawing on publicly available strategy documents published in English by universities in Australia, Canada, Italy, New Zealand, Singapore, South Africa, Sweden, and the UK as examples, I will highlight the significance of university strategy documents as an important source for understanding the interface between higher education and the SDGs. Based on the strategy documents from selected institutions in different countries, the analysis aims to shed new light on the diverse approaches for advancing sustainability in different cultural and institutional contexts. Finally, I will conclude by highlighting the significance of universities as resilient institutions—with their own communities and external partners working together to shape their sustainability priorities.

8.2 Modalities: A Lens for Linking Higher Education and Sustainability

The relationships between the SDGs and higher education have become a significant research topic in sustainability research (e.g., Chankseliani & McCowan, 2021; McCowan, 2019, and references therein). McCowan (2019) presented five modalities of higher education in connection with the SDGs, and these modalities provide a model for understanding the relationships between the sustainability goals and the diverse functions of higher education. Education and knowledge production, which are the first two modalities, correspond to the two main functions of higher education. There are some important points related to these core modalities. First, McCowan (2019) highlighted the idea that the education modality is not restricted to learning and teaching activities traditionally defined by the formal curricula—as students' participation in extra-curricular activities and work placements outside universities, for instance, can also form a significant part of their education experience. Second, the author also showed the importance of regarding knowledge production as "more fundamental than the existence of research projects, funding, administrators and publications — beyond the ranking and resources". In addition to pure and applied research, knowledge production should also encompass the increasingly common and inclusive forms of participatory research.

Community engagement and advancing values-based practices (e.g., Gibbs, 2017; Waddington, 2021) have become increasingly important components in education and knowledge production. These community and values-based practices include transformative approaches to university education (e.g., Ashwin, 2020; Jackson, 2018), and emerging forms of democratized inquiry-based activities (e.g., Chang et al., 2013; Chevalier & Buckles, 2019). As the SDGs-higher education interface continues to evolve, universities remain at the forefront in setting and shaping the wider social agenda. For instance, there is an increasing emphasis on working with underrepresented communities to promote diversity, equity, and inclusion in education and research initiatives. Online and blended approaches have contributed to the development of innovative practices in research and education, and the development of these practices has accelerated in response to the challenges and opportunities posed by the COVID-19 pandemic.

Apart from education and knowledge production, McCowan (2019) highlighted two other modalities in relation to the SDGs and external-facing functions of higher education institutions. Regarding the third modality, higher education plays an important role in "defending the principles of enquiry, deliberation and public scrutiny", thereby promoting the principle of dialogue in public engagement. This "public debate" modality is a form of intellectual and social leadership, which is perhaps best exemplified by related outreach activities. These activities may also be closely linked to education and knowledge production initiatives. In less economically developed countries and regions, higher education institutions may have significant involvement in "service provision", the fourth modality, as they may offer public services such as health care, legal advice, and engineering support as well as the provision of internet, electricity, and recreation and meeting facilities to their communities (McCowan, 2019). It is possible to broaden the concept of service provision modality to cover the wide spectrum of external-facing activities—with the goal of enriching the intellectual and cultural life of the communities that the universities serve.

The modalities of public debate and service provision therefore signify higher education's efforts to reach out to their communities by using their expertise and resources. With demonstrable societal impact, these community-facing interfaces have particular significance in relation to addressing the SDGs. In contrast, "embodiment", the fifth modality, is about higher education institutions living up to the expected standards regarding SDGs-aligned practices. Although embodiment is to a large extent an internal-facing modality as it is primarily about universities' own infrastructure and operations, the symbolic significance of 'practicing what one preaches' cannot be underestimated.

8.3 University Strategy Documents: An Important Source for Studying Higher Education Priorities in Sustainability

Higher education institutions have different focuses across the modalities—through working in partnership with their academic and non-academic community networks. These differences in institutional priorities result in a wide spectrum of approaches used for advancing the sustainability agenda. Capturing diverse institutional strategies from different types of universities around the world can therefore help provide insights into how the higher education sector addresses the SDGs in response to international trends as well as considering their own institutional, regional, and national contexts.

To academics, staff, students, and external stakeholders, institutional strategies are important documents in which high-level vision and long-term aspiration from the university leadership are articulated and communicated. The documents are product of discussions and negotiations—with shared vision and buy-in from different parts of the university communities. In other words, strategic documents are also significant cultural artifacts of institutional processes. Hinton (2012) wrote in the foreword of a guide on strategic planning for higher education leaders, highlighting the importance of the strategic planning process:

> The costs of engaging in a poor planning process range from disillusioned faculty, staff, and students, to poor use of vital resources, to failed accreditation reviews which, in turn, cause an institution to lose funding and prestige. The stakes are high, but the rewards are higher. A well designed and implemented strategic planning process can provide an institution with a forum for campus-wide conversations about important decisions. The process can also be organized to make assessment, resource allocation, and accreditation easier, and be a source of information about progress and achievement with very real meaning to those associated with the institution.

Institutional strategy documents are sometimes published on the main university website for both internal and external stakeholders. However, this open-access practice is by no means the norm. Strategic plans have vastly different targeted time frames, styles of presentation, and levels of details. In response to changes in resource priorities and senior leadership as well as emerging opportunities, strategic plans are subject to revision. It goes without saying that strategic goals, objectives, and targets do not themselves provide evidence of impact—they represent aspiration and commitment to a strategic initiative at a given time only. Despite these limitations, strategic documents offer a good window into understanding how different types of higher education institutions view their own circumstances and SDGs-related modalities. It is important to note that the absence of institutional strategies or detailed targets does not necessarily imply a lack of high-level commitment to sustainability. Nor does it imply the absence of a flourishing community with bottom-up SDGs initiatives led by academics, students, staff in collaboration with external partners.

Universities prioritize the five SDGs modalities in different ways, and the different priorities are reflected in their institutional strategies. These differences can be attributed to a range of factors, but the institution's strategic mission is an overarching consideration dictating what and how universities prioritize their resources and initiatives. For instance, large teaching-focused universities naturally have a different set of priorities compared with small institutions specializing in applied research with a well-defined disciplinary focus. Global research universities (e.g., Marginson, 2014) may have a different range of external partners to take forward core modalities (such as knowledge production and education) and community-facing modalities (including public debate and service provision) compared with development universities (e.g., McCowan, 2019), which are institutions with a focus on contributing to regional development and economic growth. The strategic missions of universities with multiple campuses located in several countries are different from those with a predominantly online presence—and these differences are closely linked to the institutional infrastructure and operations (i.e., embodiment).

Given the multiple modalities of SDGs involving cross-community collaborations in higher education, designing SDGs-aligned institutional strategies may itself be regarded as a transdisciplinary effort for university leaders. As higher education institutions do not prioritize the SDGs modalities separately but embed them in the objectives and targets, strategy documents provide unique insights into the types and range of synergies across the institutional modalities.

As a source for studying the relationships between higher education functions and sustainability, eight university-level strategies with explicit reference to the SDGs have been chosen for this exploratory study. All eight universities (Table 8.1) have significant portfolios in both research and education, with relatively high rankings in the Times Higher Education World University Rankings 2022[1] (the latest edition available at the time of writing). Despite these similarities, the universities are in different parts of the world and have their own distinct histories, traditions, and missions. Five of them appear in the SDGs-aligned Times Higher Education Impact Rankings 2021[2] (the latest edition available at the time of writing) but with significantly different positions in the league table.

This exploratory study is not meant to be a comparative study nor a comprehensive investigation—but it aims to capture a range of strategic framing of sustainability priorities for institutions with broad portfolios in both research and education. It is important to note that the selected universities are not necessarily 'typical' in their countries, nor do they represent other universities with similar missions or levels of research intensity. As the strategy documents differ significantly in length, overall scope, level of details about the implementation plan, time frames, and presentation styles, the following guiding questions are used to identify the key themes in relation to the framing of sustainability priorities:

- In what ways are the SDGs featured in the strategy?
- What are the key sustainability priorities in relation to its functions as a university?

---

[1] https://www.timeshighereducation.com/world-university-rankings/2022/world-ranking.
[2] https://www.timeshighereducation.com/impactrankings.

Table 8.1   Information about the eight universities in this exploratory study

Institution	Country	Year of establishment	Times Higher Education World University Rankings 2022[a]	Times Higher Education Impact Rankings 2021[b]
University of Auckland	New Zealand	1883	137	9
University of Bologna	Italy	1088	172	20
University of Cape Town	South Africa	1829	183	101–200
Lund University	Sweden	1666	116	–
McGill University	Canada	1821	44	101–200
Nanyang Technological University	Singapore	1991	46	–
The University of Newcastle Australia	Australia	1965	251–300	12
University of Sheffield	U.K	1905	110	–

[a]https://www.timeshighereducation.com/world-university-rankings/2022/world-ranking
[b]https://www.timeshighereducation.com/impactrankings

Key strategic priorities as presented in the institutional documents are listed for identifying common themes (see Sect. 8.4), and how these strategic priorities are framed in connection with sustainability and university functions is also highlighted for each institution.

8.4   SDGs and Sustainability Priorities in University Strategy Documents

8.4.1   University of Auckland

The university strategy is entitled "Taumata Teitei: Vision 2030 and Strategic Plan 2025" (The University of Auckland, 2020), featuring the idea of "pursuing excellence, despite uncertainty" in the Maori language. As set out in the introduction of the document, the 10-year strategy for the leading university in New Zealand has a strong focus on the institutional commitment to "ecologically sustainable systems, equitable and just society, well-being for all", which is fully in line with the SDGs. The sustainability goals are explicitly referenced in the section on 'Leading transition to sustainable ecosystems': "We will continue to be world-leading in extending the reach and significance of the Sustainable Development Goals (SDGs)."

Strategic Initiatives are structured and presented in five areas:

- Education and student experience
- Research and Innovation
- Partnerships and Engagement
- Enabling Environment
- People and Culture

There is a strong emphasis on building communities and partnerships, and promoting values in diversity, equity, and inclusion across all five areas. One of the priorities under 'Education and student experience' is on 'research-informed, trans-disciplinary, relevant and with impact for the world'—this indicates the institution's strategic commitment to advancing sustainability through developing links between research and education. Apart from developing 'ambitious research confronting humanity's greatest challenges', the institution is committed to research inspired by its geographical location in the Pacific, with impact for its communities. Under 'Partnerships and engagement', the university aims to be valued by its contributions to making 'a more sustainable future for all'. Sustainability is also highlighted with its commitment to 'achieve net-zero carbon status and to publish meaningful metrics of the University's progress towards overall sustainability' as a priority for the initiatives under 'Enabling Environment'.

8.4.2 University of Bologna

As the oldest university in continuous operation in the world, the University of Bologna "promotes sustainability as a development strategy and the Multicampus as an integration strategy" in its 'Strategic Plan 2019–2021' (Alma Mater Studiorum Università di Bologna, 2019). Mapped against one or more of the SDGs and related targets, the eight key goals in the strategy document are:

- Quality of research
- Project work
- Teaching quality
- Attractiveness
- Students
- Innovation
- Dissemination
- Sustainability

There is a correspondence between these eight goals and the five SDGs modalities presented by McCowan (2019). SDG 8 (Decent Work and Economic Growth) and SDG 9 (Industry, Innovation and Infrastructure) underpin the first two key goals, which correspond to the knowledge production modality. Although there is a focus on building capacity, funding, support, and collaborative networks for research, specific sustainability-related research themes are not highlighted in the strategy document.

The next key goals are related to education, and alignment with SDG 4 (Quality Education) is, unsurprisingly, featured in a very prominent way. SDG 17 (Partnerships for the Goals) is also referenced under 'Attractiveness', which are targets to "improve attractiveness and strengthen the international dimension of learning environments". It is therefore clear that the SDGs are used as for building the institutional internationalization strategy across education and research. 'Students' is the key goal for addressing student support and enhance their access to higher education, and several SDGs have been referenced in the university's student wellbeing, support and widening participation strategies, including SDG 3 (Good Health and Wellbeing), and SDG 10 (Reduced Inequalities).

'Innovation' and 'Dissemination' are aligned with the external-facing SDG modalities. SDG 17 (Partnerships for the Goals) and other SDGs are referenced to support the institutional goals on knowledge transfer and public engagement. Under the eighth key goal 'Sustainability', the concept of the 'embodiment' modality is reflected in the themes: promoting environmental sustainability of university buildings, social responsibility of the university and wider communities, and 're-affirming' the importance of the university's foundational values.

8.4.3 University of Cape Town

The 'UCT Environmental Sustainability Strategy' (Braune, 2020) focuses on a theme that the institution has been committed to before the launch of the SDGs. In the 'Background and Context' section, the University makes the link between environmental sustainability and the SDGs through its commitment to contributing more broadly to the SDGs.

The institutional strategy covers the following five areas:

- Learning
- Research
- Governance
- Operations
- Engagement & social responsiveness

Under 'Learning', the University focuses on developing curriculum and course contents, and aims to provide training in environmental sustainability to all students. In addition to supporting and creating links with existing research on environmental sustainability, the institution will "create a Living Lab experience via research linked to campus facilities and campus property related projects that support on-site real-life environmental sustainability research opportunities". This is evidence showing the institutional commitment to linking the research modality with the embodiment modality at the strategic level.

There is a special focus on 'Governance', under which environmental sustainability serves as a theme linking a wide spectrum of university activities, including "its facilities, human resources, student life, procurement, application of finance and general UCT culture". 'Operations' encompasses a wide range of targets, including green campus, waste, energy consumption, carbon emission, and responsible investment. Highlighting governance and operations as two separate areas in the strategy document shows the significance of the institutional structures and processes in the implementation of the strategy. Unlike the other four areas, 'Engagement & social responsiveness' is primarily about internal and external partnerships—engagement with students, faculty, staff, organizations, and the wider communities.

8.4.4 Lund University

In the first paragraph of the Introduction of the 'Sustainability Plan for Lund University, 2020–2026' (Lund University, 2020), the institution foregrounds the SDGs in its commitment to sustainability:

> As a higher education institution, regional actor and public authority, Lund University helps achieve regional, national and global objectives, including the UN's 17 Sustainable Development Goals. The Strategy for Sustainable Development at Lund University 2019–2026 clarifies our approach to understanding, explaining and improving our world and the human condition and thereby being a driving force for sustainable development.

There are two key themes in the strategy, covering the university's core functions, and how it delivers its functions as a higher education institution:

- Education, research, and external engagement
- A sustainable organization

The strategy document also presents a table showing how the institutional sustainability goals address all 17 SDGs. The goals on education, research, and external engagement address all the SDGs, whereas those linked to sustainable organization are associated with nine SDGs. This highlights the distinct contribution of higher education to advancing sustainability—addressing all the SDGs through the learning, knowledge production, and external-facing modalities. There is an emphasis on going beyond sustainability as thematic contents. For instance, under 'Education for Sustainable Development, one of the goals is on 'Sustainability perspectives to be managed as a quality issue'.

To make Lund University a more sustainable organization, the strategy sets targets with reference to institutional procedures as well as international standards in the following areas:

- HR
- Procurement
- Divestment and impact through investment
- Climate

- Waste, recycling, furnishing
- Food/conference activities
- Travel to and from work
- Premises provision and buildings
- Chemical safety

It is worth noting the targets on conference activities and chemical safety—as they reflect the institutional contexts and priorities in addressing the impact of specific university activities in relation to sustainability.

8.4.5 McGill University

'McGill University Climate & Sustainability Strategy 2020–2025' (McGill University, 2020) is built on two transversal themes: (1) climate change mitigation and adaptation, and (2) equity, diversity, and inclusion. In other words, the strategy document is both underpinned by values and has identified thematic focus within sustainability. In one of the appendices, the SDGs are referenced as they "provide a blueprint to achieve a more equitable and sustainable future for all. The SDGs represent a call to action for all nations in order to address the global challenges facing humankind. For each of the eight categories of the Strategy, the interconnections are highlighted with the relevant SDGs. As a higher education institution, McGill holds a responsibility in promoting and contributing to the achievement of the SDGs.

The eight Categories are:

- Research & Education
- Buildings & Utilities
- Waste Management
- Travel & Commuting
- Food Systems
- Procurement
- Landscapes & Ecosystems
- Community Building

The core university functions of research and education appear first but only in one combined Category. The common focus across the two functions is to "identify strategies to increase learning and research opportunities in sustainability". There are specific actions corresponding to the two functions in this Category in the strategy document. 'Applied student research' in climate and sustainability based on the campus as a living lab is an example linking knowledge production and learning modalities. It is also worth noting that indigenous-led research strategies and initiatives are mentioned in this Category. These two examples demonstrate how the two transversal themes in the strategy are embedded under the research domain.

'Community Building' is a Category that focuses on engagement and enabling processes that underpin the delivery of goals in the other Categories. The other six

Categories correspond to the embodiment modality—this clearly underscores the university's commitment to advancing sustainability through actions linked to its operations.

8.4.6 Nanyang Technological University

In the 'Sustainability Framework October 2021' (Nanyang Technological University, 2021), there is a dedicated section entitled "NTU's commitment to United Nations' Sustainable Development Goals". In this Section, the University refers to the SDGs in two ways. First, as a young research-intensive institution with high world rankings, the University showcases its SDGs-aligned research outputs: "NTU supports and promotes the principles of the Sustainable Development Goals SDG)". Based on the UN SDGs, NTU's top 5 SDG goals by publication output are below (1) Affordable and clean energy, (2) Good health and well-being, (3) Climate action, (4) Quality education and (5) Sustainable cities and communities. Second, the institution lists 13 actions spanning learning production, education, and embodiment modalities. These actions are listed under seven SDGs:

- Good health and well-being
- Quality education
- Affordable and clean energy
- Sustainable cities and communities
- Responsible consumption and production
- Climate Action
- Partnerships for the Goals

In addition to the 13 actions, the strategy identifies two key commitments linked to education and campus:

- "Inspiring and educating current and future generations about sustainability": a PhD Program in AI and Sustainability, UG common interdisciplinary core curriculum, and electives related to sustainability in curricular programs; and
- "NTU strive to achieve carbon neutrality, along with at least 50 percent reduction in carbon emissions by 2035".

There is a strong emphasis on building on the institutional strengths in research and education (e.g., flagship research centers, research domains, and pedagogical approaches), as well as setting SDGs-aligned targets on clean energy, waste, buildings etc. In contrast, public engagement and values associated with sustainability are less prominently featured in the strategic framework document.

8.4.7 University of Newcastle Australia

The SDGs are referenced as an important driver behind the development of the institutional strategy. This is evidenced by the Vice-Chancellor's opening message in the 'Environmental Sustainability Plan 2019–2025' (The University of Newcastle Australia, 2019): "Guided by the Sustainable Development Goals (SDGs) set down by the United Nations in Paris in 2015, we have created this Plan to identify our targets and priority actions to 2025. We regard our role as host to one of the UN's 17 International Training Centres for Authorities and Leaders (CIFAL) as a great honour, and we are conscious our pivotal role in promoting and increasing understanding of the SDGs begins within the grounds of our own campuses."

Apart from the SDGs, the university has also referred to the 'Talloires Declaration of University Leaders for a Sustainable Future' as an institution commitment. Despite the SDGs being highlighted as a key driver in the introduction of the strategy document, the links between the SDGs and strategic objectives are less explicit in the core strategy.

There are five areas in the strategy focusing on environmental sustainability:

- Our operations
- Engagement
- Education
- Research
- Governance

Operations appear before the core functions in research, education, and engagement. Priority actions are detailed across a wide spectrum of themes: Energy and carbon, Water, Waste and Recycling, Biodiversity and Landscaping, Environmentally Sustainable Design, Transport, Investments, and Procurement. In addition to the strategy document, the University intends to develop an institution-wide Environmental Sustainability Charter.

Regarding targets in education, the university is committed to embedding environmental sustainability across all programs and/or courses. As for research, the strategy highlights partnerships with the industry, and its flagship research centers in renewable energy and environment.

8.4.8 University of Sheffield, U.K.

In the introduction section of the 'University Sustainability Strategy 2020–2025' (University of Sheffield, 2020), the institution articulates the significance of the SDGs with a dedicated section 'Achieving the Sustainable Development Goals'. Five SDGs have been identified as particularly relevant to the strategy:

- Affordable and Clean Energy
- Climate Action

- Quality Education
- Responsible Consumption and Production
- Sustainable Cities and Communities

There is also a commitment to use the SDGs as a framework for review: "As intended by the goals, we will view sustainability holistically and will link this strategy to other activity when relevant and necessary. We will continually review our use of the SDGs to ensure that they represent the most impactful approach for the University." The institutional strategy consists of five Commitments:

- Carbon neutrality
- Research and innovation
- Education
- Campus: Biodiversity, buildings, divestment, energy, food, procurement, travel, and waste
- Our place in the city region

The 'living labs' approach—for linking research activities with the campus and city region—is mentioned under 'Research and innovation', in which external engagement with policymakers, governments, and the public is highlighted. Innovation is on knowledge transfer and research impact. The links between the university research and the Sheffield city region are articulated: "We have launched a regional sustainability programme that connects the world-class research capability in sustainability at the University with the priorities of local and regional authorities, businesses and voluntary organisations throughout the region."

The University is committed to the SDGs-aligned 'Education for Sustainable Development'. There is a focus on building on existing pedagogical and curriculum design approaches, providing faculty development and staff training, and developing student skills and graduate attributes. As part of the Commitment to 'Our place in the city region', the institution aims to "provide a unique contribution to a post Covid-19 green recovery". This shows that the University has identified and embedded sustainability for addressing current and evolving challenges in the strategy.

8.5 Discussion: Whose Priorities? Our Priorities for Advancing Our SDGs'

As illustrated by the strategy documents in this study, universities use a variety of approaches to frame priorities in sustainability for their institutions. The SDGs offer a comprehensive framework for developing strategies in diverse settings, and there are different ways to incorporate the SDGs in university strategy documents. Some universities highlight the alignment between the SDGs and their institutional mission and values, with the SDGs providing additional legitimacy to the strategic framing. Although there are examples of strategic priorities being presented under a list of relevant SDGs, the UN's goals are primarily used as a mapping framework. In other words, the SDGs do not appear to be the key driver for *prioritizing* the sustainability targets.

As revealed in this exploratory study,

- showcasing research strengths (e.g., expertise and infrastructure),
- developing education programs on sustainability (e.g., university-wide modules)
- advancing institutional values (e.g., diversity, equity, and inclusion), and
- connecting with the local communities (e.g., indigenous communities, and the city region)

are four key thematic drivers underpinning the sustainability strategies. The first two are associated with core university functions, and the last two are cross-cutting thematic drivers. There is generally a dichotomy between the priorities on core university functions (i.e., research, education, and engagement) and those associated with campus operations. How the two sets of priorities are presented varies significantly from one strategy to another. The relative importance of the two sets of priorities may therefore be an indicator reflecting the institutional priorities in sustainability.

There are a few caveats in the findings, however. Although the selected institutions operate in a range of cultural contexts and have contrasting traditions and backgrounds, they do not by any means represent a full range of framing approaches for advancing sustainability in universities. For one thing, all eight institutions have high levels of research intensity, with a relatively wide spectrum of academic disciplines represented in their institutions. It is also important to note that this group of universities does not share the same priorities with other types of universities (e.g., teaching-focused institutions, specialist universities, institutions in less economically developed regions). University strategies on sustainability without mentioning the SDGs have not been included in this study. Despite these limitations, it is possible to use the findings in this study to shed new light on 'prioritizing the sustainability agenda' in the higher education sector. Here I focus on the ownership of the strategic priorities, which is relevant to all types of higher education institutions irrespective of their missions, values, backgrounds, or functions. This consideration may help identify possible next steps for strategic enhancements—particularly timely as SDGs-aligned strategies are being implemented in different types of higher education institutions across the world.

Some of the university strategy documents in this study contain information about how the institutions intend to involve their students and other university members in the delivery of their strategic priorities. There are indeed many examples of opportunities for students, faculty, and professional services staff to shape and take part in sustainability research and education, including the widely featured 'living labs' initiatives. These are good examples of distributed leadership (e.g., Spillane, 2006)—for the implementation stages of the strategic plans. Whilst students and the wider university communities may have been involved in shaping the sustainability agenda for their institutions, the extent of their contribution is not clear. As demonstrated by the success of citizen science (e.g., Hand, 2010), distributed thinking and working in partnerships may have similarly transformative potential for universities in their efforts to advance the SDGs. Scholarship through collaborative participatory approaches (e.g., Chang et al., 2013; Chevalier & Buckles, 2019; Tong et al., 2018) is an effective way to mobilize communities to shape strategic priorities together—and these SDGs-aligned, values-based approaches may also be applied to and be embedded in the co-development of university agenda in sustainability.

Apart from university students, staff, and academics, external communities are also important partners featured in the university strategies. They include community partners in public engagement activities, industry partners for developing applied research, and policymakers for widening the societal impact of scholarship in sustainability. External connections may consist of networks of regional, national and/or international partners, and may involve mixed groups of academic and non-academic stakeholders as well. The range of external partnerships may have significant influence on the university's roles as hubs for advancing the SDGs. As observed across all the strategy documents, the dichotomy between research-education-engagement priorities and those related to campus operations may suggest opportunities for synergies between the two prioritized areas. It is likely that external partners are involved as partners in well-defined initiatives without necessarily having the knowledge about the university strategies or taking part in addressing other prioritized areas. Partnerships between a university and its external partners are therefore in the form of 'one-to-many'. If there are mechanisms and platforms for involving external partners in communities of practice for advancing the SDGs, they may help identify, initiate, and deliver strategic synergies across the less connected prioritized areas in sustainability.

In conclusion, university communities and external partners may therefore have even more significant roles to play than they are currently recognized in the university strategy documents—particularly as universities tend to have similar strategic focuses on advancing institutional values and connecting with the local communities through sustainability. For universities to be seen as exemplary organizations in advancing sustainability and the SDGs through values-based engagement with local communities, developing resilient approaches (e.g., Pinheiro et al., 2022; Shimizu & Clark, 2019) is therefore both necessary and urgent. Higher education's intellectual and academic leadership on enhancing our understanding of the interconnectivity between the SDGs is not only relevant to policymaking in the wider society but should also help inform their leaders to advance their own sustainability agenda. Community leadership across all the SDGs, including SDG 17 for "strengthening the means of implementation and revitalize the global partnership for sustainable development", is at the core of this endeavor. University strategy documents are good places to articulate the institutional vision for building even more inclusive and diverse partnerships to address the complex and diverse issues confronting humanity—together.

References

Alma Mater Studiorum Università di Bologna. (2019). *Strategic plan 2019–2021*. https://www.unibo.it/en/university/who-we-are/strategic-plan. Accessed 7 March 2022.

Ashwin, P. (2020). *Transforming university education: A Manifesto*. Bloomsbury Publishing. https://www.bloomsbury.com/uk/transforming-university-education-9781350157231/. Accessed 7 March 2022.

Braune, M. (2020). *UCT environmental sustainability strategy*. https://www.uct.ac.za/sites/def ault/files/image_tool/images/328/explore/sustainability/UCT_Environmental_Sustainability_S trategy_2021.pdf. Accessed 7 March 2022.

Chang, H., Ngunjiri, F. W., & Hernandez, K. A. C. (2013). *Collaborative autoethnography*. Routledge.

Chankseliani, M., & McCowan, T. (2021). Higher education and the sustainable development goals. *Higher Education, 81*(1), 1–8.

Chevalier, J. M., & Buckles, D. J. (2019). *Participatory action research: Theory and methods for engaged inquiry*. Routledge. https://www.routledge.com/Participatory-Action-Research-The ory-and-Methods-for-Engaged-Inquiry/Chevalier-Buckles/p/book/9781138491328. Accessed 7 March 2022.

Fumasoli, T., Barbato, G., & Turri, M. (2020). The determinants of university strategic positioning: A reappraisal of the organisation. *Higher Education, 80*, 305–334. https://doi.org/10.1007/s10 734-019-00481-6

Fumasoli, T., & Lepori, B. (2011). Patterns of strategies in Swiss higher education institutions. *Higher Education, 61*(2), 157–178.

Gibbs, P. (Ed.). (2017). *The pedagogy of compassion at the heart of higher education*. Springer. https://www.springer.com/gp/book/9783319577821. Accessed 7 March 2022.

Hand, E. (2010). Citizen science: People power. *Nature, 466*(7303), 685–687.

Hinton, K. E. (2012). *A practical guide to strategic planning in higher education* (Vol. 7). Society for College and University Planning.

Jackson, S. (Ed.). (2018). *Developing transformative spaces in higher education: Learning to transgress*. Routledge. https://www.routledge.com/Developing-Transformative-Spaces-in-Higher-Education-Learning-to-Transgress/Jackson/p/book/9780367861933. Accessed 7 March 2022.

Liu, J., Hull, V., Godfray, C. J., Tilman, D., Gleick, P., Hoff, H., Pahl-Wastl, C., Xu, Z., Chung, M. G., Sun, J., & Li, S. (2018). Nexus approaches to global sustainable development. *Nature Sustainability, 1*(9), 466–476.

Lund University. (2020). *Sustainability plan for Lund University 2020–2026*. https://www.sus tainability.lu.se/sustainability-lund-university/sustainability-strategy-and-sustainability-plan. Accessed 7 March 2022.

Marginson, S. (2014). Teaching and research in the contemporary university. In *Geoscience research and education* (pp. 11–18). Springer.

McCowan, T. (2019). *Higher education for and beyond the sustainable development goals*. Springer Nature.

McGill University. (2020). Climate & sustainability strategy 2020–2025. https://www.mcgill.ca/ sustainability/sustainability-strategy. Accessed 7 March 2022.

McGowan, P. J. K., Stewart, G. B., Long, G., & Grainger, M. J. (2018). An imperfect vision of indivisibility in the sustainable development goals. *Nature Sustainability, 2*, 43–45.

Nanyang Technological University. (2021). *Sustainability framework*. https://www.ntu.edu.sg/sus tainability. Accessed 7 March 2022.

Nature Sustainability. (n/a). *Aims & scope*. https://www.nature.com/natsustain/aims. Accessed 7 March 2022.

Nerini, F. F., Sovacool, B., Hughes, N., Cozzi, L., Cosgrave, E., Howells, M., Tavoni, M., Tomei, J., Zerriffi, H., & Milligan, B. (2019). Connecting climate action with other sustainable development goals. *Nature Sustainability, 2*, 674–680.

Pinheiro, R., Frigotto, M. L., & Young, M. (2022). *Towards resilient organizations and societies: a cross-sectoral and multi-disciplinary perspective*. Palgrave Macmillan.

Reyers, B., & Selig, E. R. (2020). Global targets that reveal the social-ecological interdependencies of sustainable development. *Nature Ecology & Evolution, 4*(8), 1011–1019.

Shimizu, M., & Clark, A. L. (2019). *Nexus of resilience and public policy in a modern risk society.* Springer Singapore.

Tong, V. C. H., Standen, A., & Sotiriou, M. (2018). *Shaping higher education with students– Ways to connect research and teaching.* UCL Press. https://www.uclpress.co.uk/products/95121. Accessed 7 March 2022.

Spillane, J. P. (2006). *Distributed leadership. The Jossey-Bass leadership library in education.* John Wiley & Sons.

The University of Auckland. (2020). *Taumata Teitei: Vision 2030 and strategic plan 2025.* https://www.auckland.ac.nz/en/about-us/about-the-university/the-university/official-publications/strategic-plan.html. Accessed 7 March 2022.

The University of Newcastle, Australia. (2019). *Environmental sustainability plan 2019–2025.* https://www.newcastle.edu.au/our-uni/sustainability. Accessed 7 March 2022.

The University of Sheffield. (2020). *University sustainability strategy 2020–2025.* https://www.sheffield.ac.uk/sustainability/strategy. Accessed 7 March 2022.

United Nations Development Programme. (2015). *The SDGS in action.* https://www.undp.org/sustainable-development-goals. Accessed 7 March 2022.

Waddington, K. (Ed.). (2021). *Towards the compassionate university: From golden thread to global impact.* Routledge. https://www.routledge.com/Towards-the-Compassionate-University-From-Golden-Thread-to-Global-Impact/Waddington/p/book/9780367341817. Accessed 7 March 2022.

Vincent C. H. Tong is the University Director of Learning and Teaching and an Associate Professor in Department of Geography and Environmental Sciences, Northumbria University. He studied Physics at Imperial College London (1994–1997) and did his doctoral research in Geophysics at Trinity College Cambridge (1997–2000). As an academic developer, Vincent is interested in research-education-outreach linkages and distributed leadership in higher education. As a physical scientist, he has led international research projects on the transdisciplinary applications of seismic imaging to oceanography, geology, astrophysics, and petroleum sciences. He has been serving as the founding Secretary of the Education Section of the American Geophysical Union since 2018.

Part V
Conclusion

Chapter 9
How Can a Resilience Approach Address SDGs?

Mika Shimizu

Abstract This Chapter, as a final chapter in this book, focuses on how the resilience approach which this book presented can address Sustainable Development Goal (SDGs) by providing the detail case analysis of the systemic challenges. Specifically, this Chapter highlights a major case featuring tradeoffs to demonstrate how the root causes of tradeoffs can be analyzed from the resilience approach point of view, and how tradeoffs can be overcome by generating synergies. Based on this and other chapters of this book, specifically to narrow the operational gaps in implementing SDGs and address the missing links of different systems/subsystems and other related components in systemic ways, specific recommendations for major stakeholders or actors are provided especially from the perspective of designing relevant projects or programs.

9.1 Overview

A resilience approach for Sustainable Development Goals (SDGs) (see Chapters 1 and 2) is not a panacea and not a totally new approach, particularly if you are only concerned with the individual parts. A resilience approach can play a major role in and around SDGs by enabling linkages of wholeness and details. In the macro and micro dimensions (scales can be different depending on contexts) of natural, human, and social systems, the emphasis can be placed on their relationships, interplays, or boundaries (i.e., "living systems") through a systemic and overarching approach to systemic risks or challenges in and around SDGs (see Chapters 1 and 3). As such, a resilience approach to SDGs paves the way for providing clues on how to address the systemic challenges in and around SDGs by identifying operational gaps and missing links in existing approaches to SDGs and linking the identified gaps and missing links to problem-solving-oriented actions.

M. Shimizu (✉)
Graduate School of Advanced Integrated Studies in Human Survivability, Kyoto University, Kyoto, Japan
e-mail: shimizu.mika.5a@kyoto-u.ac.jp

In terms of operational gaps from the macro (policy in this context) perspective, as specified in Chapter 2, although it is recognized that issues with SDGs are interlinked, at least conceptually, within international scientific and policy communities, this recognition is not necessarily translated into actions against issues such as climate change, biodiversity, and forest protection. Moreover, the understanding of the linkage tends to be limited to "issue-to-issue" linkage, although the linkage of different layers and the boundary areas, especially in "living systems" or nature-human-social systems, holds a key to implementing SDGs (see Chapters 1, 2, 4, and 5).

Based on the assessment of the current status and existing or emerging approaches, Chapter 2 proposed a resilience approach to narrow the operational gaps, which highlights the (1) linkages/boundaries across natural-human-social systems; (2) global to local linkages/boundaries; (3) linkages/boundaries among stakeholders, including among science and policy communities and local communities; (4) linkages/boundaries among different kinds of knowledge (from scientific, academic, and practical knowledge through to local or indigenous knowledge) and actions; (5) interrelations among climate change and disasters or resilience and sustainability; and (6) linkages/boundaries among various disciplines including natural, social, and human systems, sectors, and actors. Specifically, linkages/boundaries entail not only issues/risks/system linkages or boundaries, but also diverse components that enable resilience, including natural, human, and social/financial/technological resources; stakeholders; places/communities; and generations and processes across time and scales.

Regarding the emerging approaches, as discussed in Chapter 2, recent years have seen the progress of emerging approaches at the macro policy level, such as nature-based solutions (NbS) derived from the co-benefits approach, which aims to gain multi-benefits by providing overarching perspectives and approaches through ecological-social systems (IUCN, 2016). NbS originated in Europe and were later broadly adopted as an efficient approach to climate change/biodiversity in major international policy and business communities. Recently, NbS have gained traction even in international disaster policy approaches (UNDRR, 2021), although linking disaster risk management and sustainable management of nature/ecosystems was not a major policy approach in the past.

On the other hand, in observing different ongoing SDGs-related actions, it is recognized that there are many cases where related actions are causing "tradeoffs" among SDGs, i.e., among natural, human, and social systems. Since the tradeoffs are the outcome of the operational gap resulting from systemic interactions of the missing links of different systems/subsystems and other components, facing tradeoffs requires a resilience approach given the systemic challenges or the dynamics of natural-human-social systems. This approach encompasses not only the above ecological-social systems view adopted in NbS, but also overarching perspectives and approaches in natural-human-social systems, especially "living" systems (see Chapters 1 and 2). Thus, the following section highlights the tradeoffs by providing a major case featuring tradeoffs to demonstrate (a) how the root causes of tradeoffs are related to the resilience approach point of view, and (b) how tradeoffs can be overcome by generating synergies.

Especially regarding (b), the below focuses on how to identify and address missing links at the project or program level based on the understanding that the accumulation of missing links by different stakeholders can create operational gaps. As specified in Chapter 2, missing links can be differentiated from policy gaps as missing links can be identified by associating policy challenges with lessons learned, experiences and different kinds of knowledge beyond disciplines and beyond expert knowledge including local knowledge, and indigenous knowledge. Given that missing links will be identified only through relationships or collaborations, the relevant project or program (this section calls this the "micro" dimension) for implementing SDGs by stakeholders needs to incorporate perspectives or approaches relevant to the resilience approach in the project or program.

9.2   How Can the Resilience Approach Address SDGs?

9.2.1   Tradeoffs and Synergies

What is the root cause of the tradeoffs that impact SDG implementation, and how can these tradeoffs be overcome and turned into synergies? Before seeking clues on how to address these questions, it is critical to recall the essence of the resilience approach: humans, ecology or nature, and communities or societies commonly have capacities to change, adapt, recover, and transform, but these capacities can be protected, nurtured, and strengthened only if they maintain interdependent "relationships" within the nature-human-social systems, and more specifically, from the perspective that considers "natural, human, and social systems in a continuum" (see Chapter 2). Relating to this perspective, the not fixing "a dot" but "connecting dots to form lines" approach is essential in addressing the challenges, which is an overarching aspect of the resilience approach. If we fail to connect the dots, the missing link may result in tradeoffs between SDGs, i.e., an action that looks appropriate in promoting one of SDGs may negatively impact another aspect of SDGs (see Chapter 1).

Given the above, the below provides a typical case (although it is not well recognized by the public) of systemic challenges causing tradeoffs with analytic points of view from the resilience approach. The case is about the nexus of climate change, renewable energy, waste, forest/biodiversity, disasters, and community resilience.

9.2.2   Case Analysis: The Nexus of Climate Change, Renewable Energy, Waste, Forests/Biodiversity, Disasters, and Community Resilience

To combat climate change (specifically, to reduce greenhouse gasses), renewable energy is a critical policy area. According to the IPCC (2022), since 2010 there have been sustained decreases of up to 85% in the costs of solar and wind energy driven by a range of policies and laws, which have shaped the accelerated deployment of renewable energy, leading to global climate action. The 2022 IPCC statement also says that having the right policies, infrastructure, and technology in place to enable changes to our lifestyles and behavior could result in a 40–70% reduction in greenhouse gas emissions by 2050, and evidence suggests that these lifestyle changes could improve our health and wellbeing. As such, combating climate change and promoting renewable energies can provide other benefits including better health and wellbeing by creating synergies among SDGs.

However, the above does not mean that installing renewable energy generation technologies/infrastructure without conditions can contribute to combating climate change and fulfilling SDGs. Conversely, installing them *without* looking at natural, human, and social systems in a continuum or *just focusing on "a dot"* can result in tradeoffs, causing (1) large amounts of renewable technology-related waste; (2) damage to forests and biodiversity (for example, by cutting down trees to make space for the construction of solar farms); and (3) damage to lands that causes disasters such as landslides when typhoons, hurricanes, or flooding occur, in turn challenging community resilience (see (1), (2), and (3), respectively, in Diagram 9.1). But the situation can be reversed under specific conditions (see (4) in Diagram 9.1).

The focus on the above challenges here does not mean questioning the deployment of renewable energy, and it is acknowledged that we need to transition from fossil fuels to renewable energy for a sustainable future. In other words, since it is critical to seek to maximize the benefits of renewable energy technologies and minimize the adverse impacts, the below highlights the challenges to overcoming the tradeoffs that can impact the benefits of renewable technologies to transform tradeoffs into synergies, which is critical in implementing SDGs. The details for (1)–(4) are as follows (see 9.2.3 for (4) part) with analytic points of view from the resilience approach:

9.2.2.1   Waste

Recently, it has been recognized that renewable energy technologies such as solar photovoltaic (PV) panels and wind turbines that were installed decades ago were not designed according to the principles of a circular economy. It is inevitable that their waste generation will rapidly increase in the coming years (Graulich et al., 2021). For example, according to the International Renewable Energy Agency, it is estimated the solar waste will rise to 8 million tons in 2030 and 78 million tons in 2050 in the worst-case scenario. This waste may exacerbate climate change if it is not properly addressed. The waste issue is not limited to solar panels, as similar problems are looming for other renewable-energy technologies, including turbine blades and electric vehicle batteries (Atasu et al., 2021).

Circularity solutions may present an opportunity for collaboration among different stakeholders, but the current state of recycling those renewable technologies remains limited. For example, challenges in solar PV recycling include addressing the delamination, separation, and purification of the silicon from the glass; the presence of hazardous substances such as cadmium, arsenic, lead, antimony, polyvinyl fluoride, and polyvinylidene fluoride; and logistic constraints due to the height of large solar panels (Graulich et al., 2021).

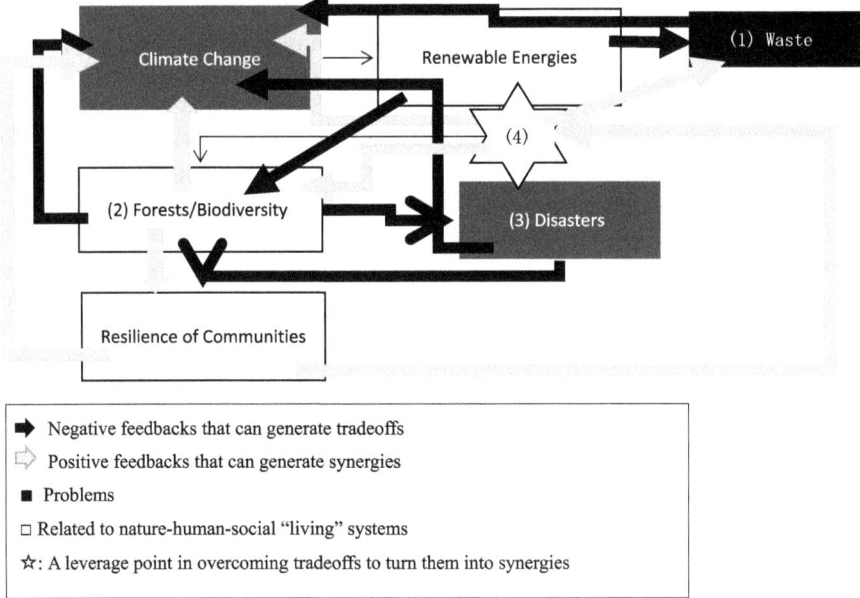

Legend:
➡ Negative feedbacks that can generate tradeoffs
⇨ Positive feedbacks that can generate synergies
■ Problems
□ Related to nature-human-social "living" systems
☆: A leverage point in overcoming tradeoffs to turn them into synergies

Diagram 9.1. Nexus of climate change, renewable energy, waste, forests/biodiversity, disasters, and community resilience

Analytic points of view from the resilience approach: While these waste issues with renewable technology infrastructures have not yet been spotlighted in combating climate change and fulfilling SDGs, just installing those infrastructures cannot solve the problem and may in fact cause damage to the environment or increase greenhouse gases. In a nutshell, this challenge requires considering "nature, human, and social systems in a continuum." Moreover, not fixing dots (e.g., one-time installation) but "connecting dots to form lines" (e.g., not only short-term but also long-term use of solar systems) is essential. Furthermore, those perspectives need to be shared among all kinds of stakeholders including the general public. Addressing renewable energy and climate change issues is not just for experts directly related to those areas, but also experts in waste and recycling, practitioners, and the general public need to be involved in co-knowledge production for better recycling and maintenance systems to make the best and longest use of solar systems.

9.2.2.2 Damage to Forests and Biodiversity

Damage to forests and biodiversity by felling trees to make space for large-scale renewable energy infrastructure, is another critical challenge, especially in terms of implementing SDGs. However, while it has been recognized that renewable energy production requires up to 10 times more land area than fossil fuel thermal facilities to produce equivalent amounts of energy, which entails *strong and often negative feedbacks between biodiversity conservation and renewable energy expansion,* policies to promote these two objectives have often been planned separately (Rehbein et al., 2020).

Relating to the above separation, the 26th UN Climate Change Conference of the Parties (COP26) in Glasgow in 2021 adopted the Glasgow Leaders' Declaration on Forests and Land Use to emphasize the critical and interdependent roles of forests, biodiversity, and sustainable land use in implementing SDGs, but did not mention the challenging feedback among climate change, biodiversity, and large-scale renewable energy infrastructure expansion. As such, although the relationships among climate change, biodiversity, and forest issues are often discussed at the policy level, it is rare for the discussions to include siting and deployment issues with large-scale renewable energy infrastructure, which may significantly impact forests and biodiversity and in turn exercerbate climate change. This may be partly because there are few assessments of the existing and near-term future renewable energy infrastructure relative to comprehensive sites for biodiversity conservation.

Regarding the above challenge, Rehbein et al. (2020) identified the extent of current and potential future overlap of renewable energy facilities and important conservation areas, showing that overlaps are numerous: Out of 12,658 large-scale renewable energy facilities distributed globally, it was found that 2206 (17.4%) currently operate inside important conservation areas. The research also pointed out that the spatial distribution of overlaps is moving from developed regions toward more biodiverse developing regions such as Southeast Asia and sub-Saharan Africa. As such, the impact of the deployment of large-scale renewable energy infrastructure on biodiversity is expanding globally.

Analytic points of view from the resilience approach: In terms of the deployment of large-scale renewable energy infrastructure, assessments of environmental impacts are essential, but how the assessments are conducted and under what kinds of governance need to be critically reviewed. Serrano et al. (2020) pointed out that assessments of environmental impacts are often funded by energy companies, often with little supervision by governments, which precludes independence, and the problem is exacerbated by the fragmentation of large projects; the absence of in-depth assessments of cumulative and synergistic environmental impacts; and decentralized administrative authority divided among the central state, regions, and municipalities. Although these kinds of management and organizational issues are sometimes disregarded in addressing SDGs, whether assessments and policy evaluations are done through the co-knowledge production, or how challenges are faced beyond the "stove-piped" specialized agencies or sectors/organizations, is a critical issue in overcoming the relevant challenges, which is a major part of resilience-based public policy (Shimizu & Clark, 2019).

9.2.2.3   Damage to Lands Causing Disasters Driven by Water-Related Hazards and Their Impacts on Communities

The typical cases for (3) in Diagram 9.1 have been seen recently in Japan: With the rapid spread of large-scale solar power generation, there has been a series of landslide disasters at facilities constructed on mountain slopes over the past few years. In 2018, when the country was hit by huge natural disasters such as torrential rains and typhoons in western Japan, a total of 57 incidents were confirmed in which panels were damaged by landslides or blown away by the wind (Mainichi Newspaper, 2021), and in 2019, the reported number of natural disasters originating from typhoons, torrential rains, etc., that occurred in the area of solar power generation facilities was 34 (METI, 2021). Although comprehensive data across years on these cases including other large-scale renewable energy facilities, are not available at present, which is another problem, these issues are emerging and evident across Japan.

Despite the lack of comprehensive data, some recent surveys suggested that this is not just a one-time issue, but will loom over the short-, mid-, and long-terms, which could damage community resilience. For example, relating to the relationships among solar panels, disaster risks, and landslides, NHK (2021) analyzed the location data of medium-scale facilities with power generation output of 500 kilowatts or more, and found that at least 1186 of the 9809 locations overlapped, even partially, with "landslide risk areas" where landslides could occur and cause damage to homes and public facilities. Moreover, the survey also found that 843 facilities overlapped, even partially, with "landslide disaster warning areas," which require evacuation and other measures when heavy rains or typhoons occur, and 249 of these facilities overlapped with "landslide disaster special warning areas," which are particularly high-risk areas.

Furthermore, the Mainichi Newspaper (2021) conducted a survey targeting 47 prefectures, and asked about the problems they were having with solar power generation. It found that the most common response was "landslides" (29 prefectures), followed by "deterioration of the landscape" (37 prefectures) and "destruction of nature" (23 prefectures).

The background behind this serious situation is "mega solar": The Feed-in Tariff (FiT) policy for renewable energy in Japan has provided incentives for the siting and construction of over 2800 new mega-solar power plants (≥ 1 MW) since its introduction in 2012. Empowered by the policy, "mega-solar" companies including the ones with foreign investments have constructed plants on mountainous areas in rural parts of Japan (Fraser & Chapman, 2018).

If a project/business comes at the expense of people's lives or well-being or natural resources, the activity is not for SDGs. Even if their original objective is to reduce green-house gases to combat climate change through increasing renewable energy, achieving one goal at the expense of other goals is against SDGs. Thus, looking at just one dot (issue) without looking at boundary areas may be led to missing links, including those among issues and stakeholders, which will in turn generate tradeoffs. Addressing SDGs need to look at common interests or common goods with all kinds of stakeholders by connecting dots to lines without sacrificing others.

Analytic points of view from the resilience approach: Those mega-solar companies are not familiar with the sites, including their geography, natural features, culture, history, and communities. Furthermore, the companies have poor communication with the local people, so the people often do not know the decision-making process for the siting. This problem structure tends to lead to unwanted constructions of mega-solar panels and unexpected natural disasters. The lack of understanding of the site, the lack of communication, and the lack of a system for involving the community members and different stakeholders in the decision-making process are the core of the challenge, which is associated with the lack of components for the resilience approach (see below).

9.2.3   Conditions to Overcome Tradeoffs and Promote Synergies

How can we overcome the challenges to renewable energy to promote synergies from the resilience approach? Based on the above analytic points of view from the resilience approach, the following five perspectives can be drawn to provide conditions to overcome tradeoffs and promote synergies (presented as conditions 1 through 9 below) in associating those with the operational foundations and pillars of the resilience approach identified in Chapter 2 (see Table 9.1).

First, not only issue to issue linkages but also boundary areas around issues are key in a problem-solving-oriented approach. Specifically, to address boundary areas, limited experts in partial areas are not enough. **Boundary areas tend to be little known or hard to grasp as a whole because the issues are complex and natural-human-social (living) systems, various stakeholders, and different contexts in regions/communities are intricately interrelated.** Thus, it is critical **to create a co-knowledge production system or scheme to synergize different kinds of knowledge including local/community/place-based knowledge through collaborations among different kinds of experts and scientists, stakeholders including local governments and businesses, and most importantly, locals/community members who may be directly or indirectly impacted. Through this process, "missing links" will be identified from different dimensions to draw insights from stakeholders (condition 1).** In this situation, it is essential for different stakeholders to **communicate well with a focus on their relationships without "stove-piped" administrations and organizations (condition 2).**

Second, given the lack of overarching assessments of the existing and near-term future renewable energy infrastructure relative to comprehensive sites for biodiversity conservation and comprehensive risks, and the lack of relevant interdisciplinary research and planning activities, it is necessary to form a process for **comprehensive assessments, including environmental impact assessments and risk assessments, in ways which are socially acceptable amongst a broad spectrum of stakeholders, as well as integrative planning and actions based on the assessments**. This can be done **by "connecting dots to form lines" or considering "natural, human, and social systems in a continuum,"** specifically **with systemic perspectives** for addressing climate change, renewable energy, waste, forests/biodiversity, disasters, and community resilience around the issue of deploying renewable energy infrastructure (**condition 3**). It is **through this process** that **implementing rigorous**

comprehensive planning and actions based on wide-ranging assessments can be possible, built upon updated knowledge from different directions (condition 4).

Third, related to the first and second points, the co-knowledge production process, specifically the **social experiment type of transdisciplinary research with an emphasis of how to co-learn and co-design with stakeholders,** is critical. As discussed in Chapter 2, a "modular, non-linear, and consistent process" that is a part of the foundation of the resilience approach is critically applicable to this stage. That is, **it is essential to structure co-learning and co-design processes in teams, including observation, prototyping, and testing or soliciting feedback through "triple-loop learning"** (Pahl-Wostl, 2009) **(condition 5).**

Unlike the incremental improvement of action strategies without questioning the underlying assumptions and a revisiting of assumptions, the essential point of the co-learning and co-design through the resilience approach is **to examine underlying structures and paradigms that go beyond the established framework to imagine various futures and contemplate their uncertainty through collaborative works with a focus on specific regional/local/community contexts. Through this process, "missing links" will be identified to induce new insights, and alternatives or combinations of different options will be drawn, which reflects specific regional/local/community contexts (condition 6).** The options may include distributed/ small-scale development that would reduce the direct environmental impacts on biodiversity, improved recycling for renewable energy infrastructure, energy efficiency, and energy-saving efforts.

Fourth, as seen from the mega-solar deployment cases in Japan, without understanding the geographical and natural features or culture and history around the construction sites of renewable energy infrastructures, the construction may end up facing a disaster, which could have a major impact on people's lives and the well-being of communities, eventually leading to an impact on community resilience in cascading ways. The root causes of such cascading impacts are the lack of respect for community members and nature and a lack of communication with the locals/community members.

Reflecting on the cases in Japan, from the perspectives of (a) linkage, (b) process, (c) temporal (time), and (d) scale, which is one of the operational pillars of the resilience approach (see Chapter 2 and Operational Pillar 2 in Table 9.1), a systemic explanation can be made as follows: The lack of (a) linkage-related components such as trust, face-to-face relations, interactive communication, and linkage between operations/fields and decision-making led to the lack of (b) process in co-knowledge production, which led to the lack of (c) multi-temporal views: short-, mid-, and long-term impacts, and (d) scale perspectives. As such, the analysis through the resilience approach articulates systemic cascading impacts from one to others. This reality emphasizes respects for community lives and nature and communication with locals/community members are critical (**condition 7**). In particular, special consideration needs to be given to vulnerable people and communities, since the cascading impacts especially affect them (**condition 8**).

Fifth, on the other hand, from the "living systems" point of view discussed in Chapter 1, communities can drive dynamics from a small scale toward wholeness (systemic dynamics), from the inside and outside of the communities, as communities may have some potential survival and self-organizing capacity—that is, "systemic capacity" (Okada, 2021). This systemic dynamics from the systemic capacity of communities can be possible if enabling environments (see Operational Pillar 5 in Table 9.1) for stakeholders to engage in processes is prioritized, which is a critical area in terms of nurturing or creating community and local resilience (Shimizu & Clark, 2019). The significance of enabling environments is discussed through case histories or studies in different chapters in this book, using alternative terms, such as "communicative spaces" by Okada in Chapter 5, and "Ba" (Japanese) setting, i.e., "formal, non-formal, and informal institutional and/or organizational settings where tacit knowing is externalized among participants, leading to the succession and modification of shared recognition and behavior" by Nakamura in Chapter 6 (**condition 9**).

Table 9.1 Association of the conditions with the operational pillars of the resilience approach

Operational pillars of the resilience approach (see details in Chapter 2)	Relevance to the conditions identified in the above case
Operational Pillar 1: "Looking at natural, human, and social systems in a continuum" / "looking at nature/ecology, different kinds of humans and communities/societies in a continuum" views	Condition 1 Condition 3
Operational Pillar 2: (1) linkage, (2) process, (3) temporal (time), and (4) scale perspectives in a balanced manner	Condition 4 Condition 7
Operational Pillar 3: Taking into account different *contexts* surrounding nature, humans, and societies, including geographical/social/economic/cultural/historical contexts, with a focus on vulnerable populations, including the aged, disabled, poor, and other vulnerable communities	Condition 6 Condition 8
Operational Pillar 4: Building/enabling a co-knowledge or collaborative knowledge production scheme or system (see more details in Chapter 2), which goes beyond the boundaries of disciplines and different kinds of knowledge and is built upon the co-learning scheme with triple-loop learning	Condition 1 Condition 5
Operational Pillar 5: "Enabling environments" as a set of interrelated conditions, such as organizational, social, environmental, and cultural conditions that are important as the environments or the interrelated conditions impact the capacity of stakeholders to engage in processes	Condition 2 Condition 9

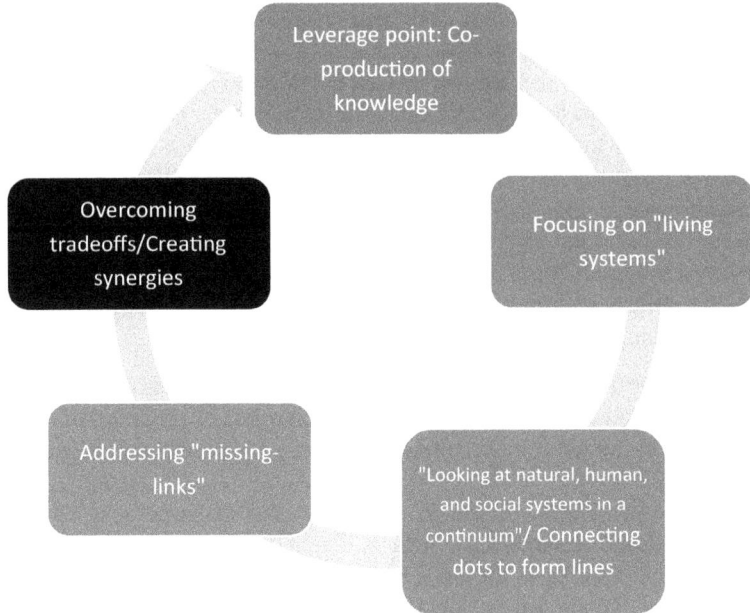

Diagram 9.2. Process cycle to overcome tradeoffs and promote synergies

Thus, the above demonstrates the overall picture in terms of how the resilience approach can address specific challenges in specific ways. While the details of the application of the resilience approach can be different depending on contexts, the essence of its application is common to different systemic challenges in and around SDGs. In sum, to overcome tradeoffs and promote synergies, creating the process cycle displayed in Diagram 9.2 is key.

9.3   Recommendations for Diverse Stakeholders Designing or Redesigning Projects/Programs for SDGs

9.3.1   Overall

The overall operational gaps in implementing SDGs identified in Chapter 2—i.e., (1) actions based on awareness of SDGs, (2) actions related to macro/micro-level linkages, and (3) actions based on attention to living systems for transformation—are critically systemic challenges. Given the systemic situations, this book proposed the resilience approach, which converges relevant systemic approaches that have been developed by combining different disciplines, experiences, and lessons learned through different kinds of systemic challenges that humanity has faced in the pursuit of a sustainable and resilient society.

Specifically, to narrow the operational gaps (in other words, gaps in implementing SDGs), all kinds of stakeholders are required to address the missing links of different systems/subsystems and other related components in systemic ways, i.e., through the co-production of knowledge with a focus of "living systems" to generate synergies of diverse knowledge, practices, and actions, as discussed above.

To design or redesign such synergy creation projects or programs for SDGs to overcome systemic challenges, it is critical for all stakeholders to check the following questions from the beginning (prototype) stage of the design or redesign stages of a project or program:

a. How can the project or program address systemic challenges in and around SDGs, or common goods, and what is the goal of the project or program, in consideration of natural-human-social changing environments, potential disturbances, and uncertainties?

b. What kind of synergies can the project or program generate in systemic ways; that is, from small- to large-scale impacts, and in the near-, mid-, and long- terms?

c. To achieve (a) and (b), what kinds of learning processes will be taken considering "natural, human, and social systems in a continuum" or "nature or ecology, different kinds of humans, communities and societies in a continuum" or "connecting dots to form lines," and in collaboration with whom?

d. How will the design or redesign of project or program grasp the contexts of the sites or fields in which the projects or programs are directly or indirectly engaged?

e. How will the project or program address vulnerable people and communities so as not to leave them behind?

f. Who can play a coordinating or "middleman" role in operating the project or program by looking at both the details and the whole in a continuum?

g. How will the project or program address linkages between actors, organizations, or sectors; between operations/fields and decision-making; and among resources and coordinating functions? What kinds of communication will be used to make linkages possible?

h. What kinds of "enabling environments" will the project or program provide for all stakeholders to engage in building the co-knowledge production/collaborative knowledge production system?

Based on the above premises, specific recommendations for major stakeholders or actors, distilled from this and other chapters of this book, are provided in the following subsections.

9.3.2 Policy Community/Decision Makers/Governments

As discussed in Chapters 1 and 4, systemic changing environments related to natural-human-social systems, risks in and around SDGs, and deep uncertainty pose structural challenges to public policy. The changes necessitate a paradigm shift from the traditional top-down, linear, and "silo" approach (seeing only pieces and not the whole) to actions linking the macro–micro levels, including policy-science-community linkages and co-learning and co-knowledge production based approaches. In this paradigm shift, it should be noted that the interrelated conditions of organizational, social, and cultural elements impact the capacity of stakeholders to engage in processes related to SDGs. Therefore, the traditional type of rigid, fragmentized policy formation may lead to negative feedbacks on their capacities, which in turn could be a cause of stagnation in undergoing the transformation necessary to implement SDGs.

Thus, to enable the paradigm shift, the policy community, decision makers, and governments need to engage in resilience-based public policy (Shimizu & Clark, 2019), which promotes co-production of knowledge through different kinds of knowledge inside and outside of policy systems with stakeholders' systemic capacities for systemic dynamics for transformation. The resilience-based public policy also allows comprehensive assessments in which a broad spectrum of stakeholders are engaged, as well as integrative planning and actions based on assessments, linking interfaces amongst policy, science, and communities.

As such, the policy community, decision makers, and governments need to play a role in creating schemes/relevant systems/spaces or enabling environments for this type of policy formation based on co-knowledge production beyond sectional boundaries. To play this role, rather than just a minor system change, they need to see the systemic aspects of the whole picture not only outside of the policy system but also inside of it, including how to cultivate human systemic capacities and create enabling environments for the paradigm shift.

9.3.3 Research Institutions/Scientists/Researchers

Given the characteristics of the resilience approach to SDGs, as specified in Chapters 1 and 2 and the above, there is a need for a social experiment type of transdisciplinary research with an emphasis on how to co-learn and co-design with stakeholders to address systemic challenges in and around SDGs. Research institutions/universities/scientists or researchers need to play a major role in initiating such transdisciplinary research. Although the term "transdisciplinary research" tends to be used as a buzz word, it is necessary to spotlight its essence: (1) understanding the complexity of the problem; (2) emphasizing the diversity of perspectives on a problem, and linking different types of knowledge through discourse; and (3) addressing a common good (Renn, 2021).

To structure, formulate, and operationalize this type of transdisciplinary research, it is essential for research institutions/universities/scientists or researchers to incorporate the foundations and operational pillars of the resilience approach specified in Chapter 2 into their projects or programs for SDGs, based on understanding the systemic risks and deep uncertainty as underlying issues, as well as the significance of macro–micro linkages and narratives, as discussed in Chapters 3–4.

9.3.4 Educational Institutions/Educators

Specifically for educators, including university researchers who engage in university educations as well, since nurturing human capacities for learning and designing the adaptive capacity to deal with systemic challenges is a leverage point for our sustainable and resilient society, education needs to a play a role in developing such human capacities through educational projects or programs. For this, as Noguchi points out in Chapter 7, Education for Sustainable Development (ESD) incorporates "co-learning and co-creation of knowledge in an adaptive approach" as a core philosophy, which is also addressed in SDG Target 4.7. As such, while the core philosophy of ESD is considered as just one of the 17 goals or 169 targets, "the core philosophy holds a key in achieving all SDGs." Based on this philosophical platform, educators can develop specific projects or programs for nurturing human capacities for learning and designing the adaptive capacity of systems through nature-human-social linkages to deal with systemic challenges using the resilience approach. There are three major points in developing relevant projects or programs to note:

1. Students need to have opportunities to learn what systemic challenges are in nature-human-social systems. Although there are many sources to learn about specific issues separately, such as climate change, biodiversity, or disasters, generally speaking, there are few educational projects or programs that focus on the linkages of those issues, or more specifically how they are systemically linked.
2. Related to the above point, learning about systemic issues cannot take place solely at a computer or desk. Students need to learn the challenges through direct experiences to draw insights and apply them to their own issues, even on a small scale. As an alternative to direct or physical experiences, "narratives," which were discussed by Chabay in Chapter 4, can play a role in that they are a communication method rooted in the field, connecting events and things to form lines to draw responses. As such, the essence of narratives is closely related to the resilience approach.
3. To make a project or program based on (1) and (2), as Tong stressed in Chapter 8, universities or other educational institutions or university researchers/educators need to have mechanisms and platforms for involving external partners in communities of practice for advancing SDGs to deliver synergies across the less connected or prioritized areas of SDGs.

Regarding the above points 1–3, as a prototype educational program, the author designed a project called "Looking at Nature, Society and Humans in a Continuum in Yakushima" in 2020–2021. It was implemented as a special program of the UNESCO Chair by Kyoto University, targeted at graduate/undergraduate students across schools, funded by the Ministry of Education, Culture, Sports, Science and Technology. The underlying theme of the special program was "the goals in SDGs do not exist separately but are interconnected where nature and human society are interlinked. The challenges in and around SDGs are caused by activities in human society, and at the same time bring about systemic impacts on nature and human society in the short, mid-, and long-terms extensively on a broad scale."

Under this theme, the special program aimed at allowing students to experience what it means to grasp "nature, humans and societies in a continuum" in the field in Yakushima. Yakushima in Kagoshima Prefecture, Japan is a World Heritage designated by UNESCO. It was chosen as a place for learning not only about diversity and the beauty of nature, but also human-nature relationships and nature-human-social linkages, their details and the whole, and those boundaries. Specifically, through collaborations with scholars and practitioners from different disciplines including engineering, water science, forestry, public policy, sociology, arts; local governments; communities; and the forest industry, the special program provided learning with a focus on (i) lowering the "walls" within and around us; (ii) grasping linkages among nature, society, and humans; (iii) reflecting on ourselves; and (iv) connecting insights drawn from the program (which focuses on "looking at trees and forests in a continuum") to how we approach SDGs. As a guidance for future work, the project also produced a booklet drawn from this project through narrative and visual methods ("narrative visual book") for addressing a sustainable society (the final report can be seen in: https://resilienceinitiative.com/wordpress/wp-content/uploads/final-report.pdf).

In order to enable this kind of transdisciplinary educational program, as noted by Tong in Chapter 8, collaboration between educational institutions and external partners is key, and it needs to be prioritized to structure mechanisms and platforms through the collaboration to advance SDGs. Universities in particular need to take the lead in developing resilient approaches to implementing SDGs, as seen in the above example.

9.3.5 Communities/Innovators/Individuals

As noted above and in Chapter 1, given that communities are drivers for dynamics from a small scale toward wholeness (systemic dynamics) for transformation, and dynamics can be possible through self-organizing capacity, it is critical for innovators or individuals who recognize the significance of systemic dynamics to lead the way in nurturing communities and working on the systemic challenges faced by communities collaboratively, while paying attention to the following points:

1. As a baseline in nurturing communities and promoting collaboration, community leaders and members need to share the idea of what are their "commons," which belong not to limited interests but common interests, and moreover, can be nurtured for community members by community members to relay to the next generations.

2. In nurturing the commons and communities, it is critical to create a "co-production knowledge" space (in other words, "communicative space" in Chapter 5 and "Ba" in Chapter 6) and relevant processes, by emphasizing the relationships not only inside the community but also outside of it, and the resources (human networks and skills, forests, seas, landscapes, indigenous/local knowledge, etc.) shared through the relationships.

3. Based on the understanding of the "commons" among community members and through the above space and process, it is critical to review ideas for community projects related to SDGs through multiple lenses, including (a) linkages (e.g., trust among people, different knowledge, resources, and inside and outside of the community), (b) process (e.g., assessment-learning and community members' participation), (c) time (e.g., short to long term), and (d) scale (e.g., scaling both in and up the work scope).

4. Through the review process, it is important to identify missing links that can be drawn through insights from the co-knowledge production process, by looking at nature (ecology), humans, and society in a continuum, and paying attention to vulnerable people.

5. Based on the missing links identified and shared by community members, it is critical to (re)combine the available resources and (re)structure the program or project design.

9.4 Conclusion

Based on the chapters in this book, this concluding chapter articulated how the resilience approach can address SDGs from different dimensions and scales. It presented a major case analysis focusing on how tradeoffs can be overcome by drawing synergies, specifically how to identify and address missing links at the project or program level.

Thus, this book demonstrated how approaches to systemic challenges in and around SDGs can impact the trajectory of actions, processes in actions, and in turn, outcomes from actions. The proposed resilience approach to SDGs is targeted at creating a trajectory for better processes and outcomes, with a focus on linkages of wholeness and details or the macro and micro dimensions (scales can be different depending on contexts) of natural, human, and social systems. The emphasis is on their relationships, interplays, or boundaries, that is, living systems, paying attention to the missing links.

The resilience approach can be incorporated into any activity related to systemic challenges in and around SDGs at any level from policy to community, or even individual activities. Every step in implementing SDGs, even a small step, is rooted in the ground presented by the resilience approach and can generate synergistically positive feedbacks, and in turn, collective positive impacts on natural-human-social systems. This will direct human society toward transformation to promote the implementation of SDGs.

Having said that, it should be noted that the resilience approach cannot be "grafted" on to any activity if it does not have the capacity to absorb or adapt to systemic changes and to transform us for the sake of better relationships in natural-human-social systems or nature-humanity-society. Thus, while working on SDGs, we need to engage in better capacity building of ourselves by creating common grounds or spaces rooted in the resilience approach.

Although this book could not cover how the resilience approach can apply to businesses which work on SDGs and to specific cases for developing countries, the resilience approach can be applied to different contexts and scales, especially because the approach addresses the roots or essence of linkages and boundaries in nature-human-social living systems by emphasizing the details and the whole in a continuum. While specific analysis for cases for businesses or developing countries through the points of views of the resilience approach needs to be done through future publications, the ongoing research activities by the author or colleagues present some signs of applicability for the relevant cases to pave the way for problem-solving oriented approaches.

From a broader perspective, this book attempted to build linkages from (1) the global to local levels, (2) sustainability to disaster management, and (3) policy-science-communities, where many missing links exist, often preventing the implementation of SDGs. Identifying missing links cannot be done automatically, and it is difficult to identify them in a fixed group or committee of members, or through a fixed process and scale. As the resilience approach proposes looking at natural-human-social systems in a continuum, review activities must be at least linked to linkage, process, time, and scale perspectives, to build co-knowledge production systems; in other words, it is imperative for us to review our society through different axis to gain new insights and synergies to seek viable solutions. Thus, humans need to make the most of our creativity and imagination toward acceleration to SDGs.

References

Atasu, A., Duran, S., & Van Wassenhove, L. N. (2021). *The dark side of solar power. Sustainable business practices.* https://hbr.org/2021/06/the-dark-side-of-solar-power. Accessed 5 April 2022.
Fraser, T., & Chapman, A. J. (2018). Social equity impacts in Japan's mega-solar siting process. *Energy for Sustainable Development, 42,* 136–151.

Graulich K., Bulac, W., Betz, J., Dolega, P., Hermann, C., Manhart, A., Bilsen, V., Bley, F., Watkins, E., & Stainforth, T. (2021). *Emerging waste streams—Challenges and opportunities*. https://www.oeko.de/fileadmin/oekodoc/EEA_emerging-waste-streams_final-report.pdf. Accessed 30 April 2022.

Intergovernmental Panel on Climate Change. (April 4, 2022). IPCC PRESS RELEASE: The evidence is clear: The time for action is now. We can halve emissions by 2030. https://www.ipcc.ch/site/assets/uploads/2022/04/IPCC_AR6_WGIII_PressRelease_English.pdf Accessed 5 April 2022.

IUCN. (2016). *A resolution at the 2016 World Conservation Congress (WCC-2016-Res-069)*. https://portals.iucn.org/library/sites/library/files/resreex.iles/WCC_2016_RES_069_EN.pdf. Accessed 22 February 2022.

Japan Broadcasting Corporation (NHK). (July 18, 2021). *Analysis of solar power facility siting over 1100 sites at risk of landslides*. (Japanese). https://www3.nhk.or.jp/news/html/20210718/k10013 145161000.html. Accessed 20 April 2022.

Ministry of Economy, Trade and Industry (METI). (2021). *Response to recent natural disasters involving electrical facilities*. (Japanese). https://www.meti.go.jp/shingikai/sankoshin/hoan_s hohi/denryoku_anzen/pdf/025_02_00.pdf. Accessed 20 April 2022.

Okada, N. (2021). *Another challenge: Systemic thinking and design for community-based/humans-focused disaster risk governance*. Invited speech a DPRI SOGO-Bosai seminar 50th Session on 26th November, 2021.

Pahl-Wostl, C. (2009). A conceptual framework for analysing adaptive capacity and multi-level learning processes in resource governance regimes. *Global Environmental Change, 19*(3), 354–365.

Rehbein, J. A., Watson, J. E., Lane, J. L., Sonter, L. J., Venter, O., Atkinson, S. C., & Allan, J. R. (2020). Renewable energy development threatens many globally important biodiversity areas. *Global Change Biology, 26*(5), 3040–3051.

Renn, O. (2021). Transdisciplinarity: Synthesis towards a modular approach. *Futures, 130*(April), 102744.

Serrano, D., Margalida, A., Pérez-García, J. M., Juste, J., Traba, J., Valera, F., Carrete, M., Aihartza, J., Real, J., Mañosa, S., Flaquer, C., Garin, I., Morales, M. B., Tomás Alcalde, J., Arroyo, B., Sánchez-Zapata, J. A., Blanco, G., Negro, J. J., Tella, J. L., … & Donázar, J. A. (2020). Renewables in Spain threaten biodiversity. *Science, 370*(6522), 1282–1283.

Shimizu, M., & Clark, A. L. (2019). *Nexus of resilience and public policy in a modern risk society*. Springer.

The Mainichi Newspaper. (June 28, 2021). *Solar power generation is "pollution" (part 1 and 2): Destruction of nature and deterioration of landscape trouble in 37 prefectures*. (Japanese). https://mainichi.jp/articles/20210628/ddm/001/040/152000c. Accessed 25 April 2022.

United Nations Office for Disaster Risk Reduction (UNDRR). (2021). *Words into action: Nature-based solutions for disaster risk reduction*. https://www.undrr.org/words-action-nature-based-sol utions-disaster-risk-reduction. Accessed 11 February 2022.

Mika Shimizu is an Associate Professor in Graduate School of Advanced Integrated Studies in Human Survivability, Kyoto University. Her long years' experiences as a policy researcher in East-West Center in Washington DC and Honolulu, Hawaii in the United States greatly contributed to publishing this book. She holds an M.A. from American University and a Ph.D. in International Public Policy from Osaka University (2006). She has been extensively involved in interdisciplinary and transdisciplinary research projects related to disasters/infectious diseases, sustainability, and climate change issues with the focus on resilience. Her major publications include Nexus of Resilience and Public Policy in a Modern Risk Society (Co-Author: Allen Clark, Springer, 2019).